High Performance and Optimum Design of Structures and Materials V

Encompassing Shock and Impact Loading

WITPRESS

WIT Press publishes leading books in Science and Technology.
Visit our website for the current list of titles.
www.witpress.com

WITeLibrary

Home of the Transactions of the Wessex Institute.
Papers contained in this volume are archived in the WIT eLibrary in volume 209 of WIT
Transactions on the Built Environment (ISSN 1743-3509).
The WIT eLibrary provides the international scientific community with immediate and
permanent access to individual papers presented at WIT conferences.
Visit the WIT eLibrary at www.witpress.com.

INTERNATIONAL CONFERENCE ON HIGH PERFORMANCE AND OPTIMUM
STRUCTURES AND MATERIALS ENCOMPASSING SHOCK AND IMPACT
LOADING

HPSM/OPTI/SUSI 2022

CONFERENCE CHAIRMEN

Santiago Hernández
University of A Coruña, Spain
Member of WIT Board of Directors

Graham Schelyer
University of Liverpool, UK

INTERNATIONAL SCIENTIFIC ADVISORY COMMITTEE

ORGANISED BY

Wessex Institute, UK
University of A Coruña, Spain
University of Liverpool, UK

SPONSORED BY

WIT Transactions on the Built Environment

*International Journal of Computational Methods
and Experimental Measurements*

WIT Transactions

Wessex Institute
Ashurst Lodge, Ashurst
Southampton SO40 7AA, UK

We would like to express thanks to all the conference Chairs and members of the International Scientific Advisory Committees for their efforts during the 2022 conference season.

Conference Chairs

High Performance and Optimum Design of Structures and Materials V

Encompassing Shock and Impact Loading

Editors

Santiago Hernández
University of A Coruña, Spain
Member of WIT Board of Directors

Graham Schelyer
University of Liverpool, UK

WITPRESS Southampton, Boston

Editors:

Santiago Hernández
University of A Coruña, Spain
Member of WIT Board of Directors

Graham Schelyer
University of Liverpool, UK

Published by

WIT Press
Ashurst Lodge, Ashurst, Southampton, SO40 7AA, UK
Tel: 44 (0) 238 029 3223; Fax: 44 (0) 238 029 2853
E-Mail: witpress@witpress.com
http://www.witpress.com

For USA, Canada and Mexico

Computational Mechanics International Inc
25 Bridge Street, Billerica, MA 01821, USA
Tel: 978 667 5841; Fax: 978 667 7582
E-Mail: infousa@witpress.com
http://www.witpress.com

British Library Cataloguing-in-Publication Data

A Catalogue record for this book is available
from the British Library

ISBN: 978-1-78466-471-8
eISBN: 978-1-78466-472-5
ISSN: 1746-4498 (print)
ISSN: 1743-3509 (on-line)

The texts of the papers in this volume were set individually by the authors or under their supervision. Only minor corrections to the text may have been carried out by the publisher.

Preface

This volume contains a selection of the contributions presented at the 2022 International Conference on High Performance and Optimum Structures and Materials Encompassing Shock and Impact Loading. The event was the merging of the International Conference on High Performance and Optimum Design of Structures and Materials that started in Southampton (UK) in 1989 and the International Conference on Structures under Shock and Impact that started in Cambridge (USA) also in 1989. The new forum combines the objectives of both previous individual conferences.

The 2022 edition of the conference was organized by The Wessex Institute and the University of Liverpool and the venue selected was Lisbon, the beautiful city capital of Portugal. Nevertheless, the health crisis created by the COVID-19 that continued reducing people´s mobility and the difficult political and military situation in Eastern Europe forced the transformation of the in person conference into an online forum.

The use of novel materials and new structural concepts nowadays is not restricted to highly technical areas like aerospace, aeronautical applications or the automotive industry, but affects all engineering fields including those such as civil engineering and architecture. The conference addressed issues involving advanced types of structures, particularly those based on new concepts. Contributions highlighted the latest development in design and manufacturing issues.

Most high-performance structures require the development of a generation of new materials, which can more easily resist a range of external stimuli or react in a non-conventional manner. Particular emphasis deserves intelligent structures and materials as well as the application of computational methods for their modelling, control and management.

The conference addressed the topic of design optimisation, this numerical technique have much to offer to those involved in the design of new industrial products, as the appearance of powerful commercial computer codes has created a fertile field for the incorporation of optimisation in the design process in all engineering disciplines.

The performance of the structures under shock and impact loads was another objective of the meeting. The increasing need to protect civilian infrastructure and industrial facilities against unintentional loads arising from accidental impact and explosion events as well as terrorist attacks is reflected in the sustained interest worldwide. While advances have been made in the last decades, many challenges remain, such as developing more effective and efficient blast and impact mitigation approaches than those that currently exist or assessing the uncertainties associated with large and small scale testing and validation of numerical and analytical models. All of that aimed to a better understanding of critical issues relating to the testing behaviour, modelling and analyses of protective structures against blast and impact loading.

The Conference tried to bring together experts in fields of materials definition and characterization, structural designers, architects, mechanical and civil engineers and experts working in experimental facilities or computing companies creating software for the fields of the conference.

Contributions to new materials included nanomaterials, high performance concrete and advanced composites. Papers on optimization techniques contained applications to timber structures, cable stayed bridges with different perspectives and enhanced approaches for topology optimization. Several presentations were devoted to innovative concrete structures showing the results of real scale test or computer modelling. The topic of blast and impact loads attracted papers on the influence of the shape of charge load, affects of impacts in aerospace and concrete structures or hydrodynamic impact.

The event attracted researchers from many countries and the scheme of the online event arranged by the Wessex Institute made it easy for the participants to access to the papers accepted after a rigorous peer review procedure. The Q&A sessions provided a friendly atmosphere and were very useful for interaction between authors and delegates and interchange of new ideas and initiatives.

The papers – as with others presented at Wessex Institute conferences – are part of the WIT Transactions in Engineering Sciences series and are archived online in the WIT eLibrary (www. witpress.com/elibrary), where they are freely available to the international scientific community.

We consider this volume to be a valuable addition to the existing literature on the subject and will be of interest to any researcher working in these scientific fields.

The Editors are grateful to the authors for their contributions and to the members of the International Scientific Advisory Committee of the conference and to the reviewers for their help in ensuring the high quality of the contents of this text.

The Editors, 2022

Contents

SECTION 1
INNOVATIVE MATERIALS
AND PRODUCTS

SELECTIVE REINFORCEMENT OF JOINING INTERFACE USING NANOFIBERS IN SINGLE-LAP JOINTS OF THERMOPLASTIC COMPOSITES FABRICATED BY THE INJECTION OVERMOLDING PROCESS: CREEP DEFORMATION BEHAVIOUR

KOKI MATSUMOTO[1], MASAYA ITABASHI[1], AKIRA KAWASUMI[1],
KENICHI TAKEMURA[1] & TATSUYA TANAKA[2]
[1]Department of Mechanical Engineering, Kanagawa University, Japan
[2]Department of Mechanical Engineering and Science, Doshisha University, Japan

ABSTRACT
An injection overmolding process enables molding and welding at the same time: a discontinuous fiber-reinforced thermoplastic is injected onto the thermoformed continuous fiber-reinforced thermoplastic composites for the fabrication of complex shape parts, namely, ribs and bosses. Since the joining strength is significantly influenced by process parameters, such as resin temperature and molding pressure during the overmolding process, achieving reliable joining strength is important for increasing the load bearing capacity. The nanofibers have great potential to increase the toughness of fiber reinforced composites as secondary reinforcement. Furthermore, selective reinforcement is allowed by nanofiber addition in the matrix onto the fiber surface or interlaminar region of laminated composites. Thus, we previously proposed the selective addition of nanofillers at the joining interfaces to increase the joining strength. In this study, we attempt to reveal the effect of cellulose nanofiber (CNF) addition on creep properties for long-term use under constant load. The shear creep test was conducted under various loads and various temperatures using a self-designed fixture. Furthermore, the debonded surface of a single lap joint was observed by optical microscopy and scanning electron microscopy. We discovered that 1.0 wt% CNF addition increased the creep failure time and decreased the creep strain at the same load. Furthermore, the creep rate was significantly decreased by CNF addition regardless of temperature.
Keywords: creep deformation, injection overmolding process, cellulose nanofibers, single lap joint.

1 INTRODUCTION
In recent years, multiple materials, such as metals and fiber reinforced plastic composites (FRPs), have been used instead of single materials in the automotive industry [1]. The multimaterial concepts aimed to acquire both benefits of materials while reducing the weight. Moreover, numerical algorithms of topology optimization have been developed to determine the material selection, location and connectivity of various materials [2], [3]. To join different material components, mechanical fasteners, such as riveting and bolting, require drilling holes in FRPs [4]. However, since the predrilling hole leads to deterioration of the mechanical performance of FRPs [5], adhesive bonding is adopted even though the curing of adhesives and pretreatments of metal surfaces require excessive time. While thermosets can be joined only by mechanical fastening methods and adhesive bonding, thermoplastics can be thermally joined by welding-based technologies [6]. To date, some welding technologies (such as ultrasonic welding, laser welding, induction welding, and friction welding) have been suggested for joining carbon fiber reinforced thermoplastics (CFRTPs) and metals [4].

Furthermore, the CFRTP component requires a high shape flexibility, high mechanical properties and a productive manufacturing process for joining metals. Conventionally, short fiber- or long fiber-reinforced thermoplastics have been used for molding products fabricated by injection molding. While injection molding enables the fabrication of complex parts, the mechanical properties of discontinuous fiber-reinforced thermoplastics (DiCoFRTPs) are inferior to those of continuous fiber-reinforced thermoplastics (CoFRTPs). Conversely, the shape flexibility of CoFRTP is limited compared with that of DiCoFRTP. Hence, an overmolding process, in which the DiCoFRTP are molded onto the substrate of CoFRTP, has great potential to fabricate the complex shape parts with high mechanical properties [7], [8]. In particular, the injection overmolding process enables fast welding and fast molding at the same time by thermoforming the CoFRTP substrate while closing the mold and overmolding the polymer melts onto the substrate to form a rib and boss. Furthermore, overmolding by additive manufacturing onto a CoFRTP substrate using a 3D printing technique has also been proposed [8].

However, the joining strength between the substrate and the overmolded part is strongly influenced by molding conditions, namely, pressure and temperature conditions during overmolding. The welding mechanisms are described by (i) intimate contact development by providing the pressure and (ii) interdiffusion of polymer chains across the interface by heat transfer [9]. Furthermore, the crystalline structure of semicrystalline polymers, such as polypropylene (PP), polyamide (PA) and polyether ether ketone (PEEK), strongly depends on the temperature-related processing parameters during conventional injection molding [10]. Thus, the joining strength of the overmolded hybrid composite parts is controlled by the crystalline state and molding conditions. However, to further improve the joining strength, the joining interface should be reinforced directly.

Recently, there have been attempts to introduce nanofillers into conventional composites as secondary fillers, especially for thermoset composites. The nanofillers enable the selective reinforcement of composites at the (i) matrix, (ii) interface between the fiber and matrix, and (iii) interlaminar region of laminates [11]. Furthermore, the addition of carbon nanotubes into epoxy adhesive improved the fatigue life of single lap joints [12]. Other nanofillers have been incorporated into adhesives to improve interfacial interactions [7]. Thus, we previously attempted to use nanofillers to reinforce the joining interface in the overmolding process [13], [14]. By interleaving the optimum loading of nanofiller-filled polymer films between the CoFRTP laminate substrate and injected polymer, the interfacial laminar shear strength increased by 52% [13], and the lap shear strength increased by 32% [14]. However, for the long-term use of overmolded hybrid composites, the fatigue and creep properties should be considered to ensure the reliability of the joining strength.

In this study, the effect of cellulose nanofiber (CNF) addition at the interface between the CoFRTP substrate and injected polymer of a single lap joint is newly discussed from the viewpoint of creep properties. To discuss the effect of CNFs on the creep properties, the applied load and temperature were varied, and the debonded surface was observed by optical microscopy (OM) and field emission scanning electron microscopy (FE-SEM).

2 EXPERIMENTAL PROCEDURES

2.1 Materials and fabrication process of single lap joints

Unidirectional (UD) tape (TAFNEX®), in which PP is impregnated into unidirectional continuous CF fibers, is provided by Mitsui Chemicals, Inc., Japan. The volume fraction of CF is 50%–70%, and the thickness of UD tape is approximately 150 μm. For the injection polymer and the polymer matrix of the nanocomposite film at the joining area, PP (J107G, Prime Polymer Co. Ltd., Japan) was used. In this study, to discuss the effect of CNF addition, pure PP is used for polymer injection. The CNFs (WFo-10005, BiNFi-s) were purchased from Sugino Machine Co., Ltd, Japan. The average diameter and length of the CNFs were approximately 10–50 nm and 1–2 μm, respectively. The specific surface area is 120 m^2/g.

The details of the fabrication process of a single lap joint reinforced by CNFs at the joining area are described in the literature [14]. The fabrication processes are divided into two steps: (A) fabrication of CFRTP laminate and (B) overmolding of PP onto the CFRTP substrate. In process (A), 20 sheets of UD tape were stacked as cross-ply laminates, and CNF-filled PP nanocomposite films were placed at the joining side. The direction of CF at the joining interface was the load direction of the creep test. The stacks were pressed through a heat press with water cooling channels (MP-WCL, Toyo Seiki Seisaku-Sho, Ltd., Japan) at 190°C, 2.0 MPa of hydraulic pressure for 10 min to achieve a thickness of 3.0±0.2 mm. After that, the heat-pressed laminates were cooled at a cooling rate of 20°C/min. The laminates were cut into rectangular specimens 50 mm in length and 10 mm in width.

In process (B), pure PP was overmolded onto CFRTP laminates through an injection molding machine (EC5P-0.1B, Shibaura Machine, Co., Ltd., Japan) with a maximum clamping force of 50 kN. The cylinder temperature and injection speed were varied during the injection overmolding process, while the other conditions were fixed, as presented in Table 1. During injection overmolding, the pressure and temperature data at the joining section were measured by a pressure sensor (SSE series, Futaba Corporation, Japan) and temperature sensor (EPSSZL series, Futaba Corporation, Japan), respectively. The used temperature sensor could detect the resin temperature with high responsiveness in 8 ms by the optical fiber infrared method. Finally, the single lap joint has a joining area of 12.5 mm in length and 10 mm in width, as shown in Fig. 1(a).

Table 1: Molding conditions of the injection overmolding process.

Parameter	Set value		
Screw rotational speed (min^{-1})	100		
Injection speed (mm/s)	40		80
Back pressure (MPa)	2.0		
Cylinder temperature (°C)	230	240	250
Mold temperature (°C)	60		
Holding pressure (MPa)	40 (1st)		10 (2nd)
Holding time (s)	15 (1st)		5 (2nd)
Cooling time (s)	20		

2.2 Tensile shear test

The lap shear strength was evaluated by using a tensile shear test with a universal testing machine (AG-IS, Shimadzu Corporation, Japan). As mentioned in Pisanu et al. [15], the conventional tensile shear test could not conduct the pure shear test since the bending moments are generated during the tensile shear test of a single lap joint. Thus, the self-designed fixture was used, as shown in Fig. 1(b). To prevent the bending moment during the tensile shear test, side plates were adopted, as shown in Fig. 1(c). The tensile shear test was repeated five times for each specimen at a tensile speed of 1.0 mm/min at room temperature. The lap shear strength was obtained by dividing the maximum load by the joining area.

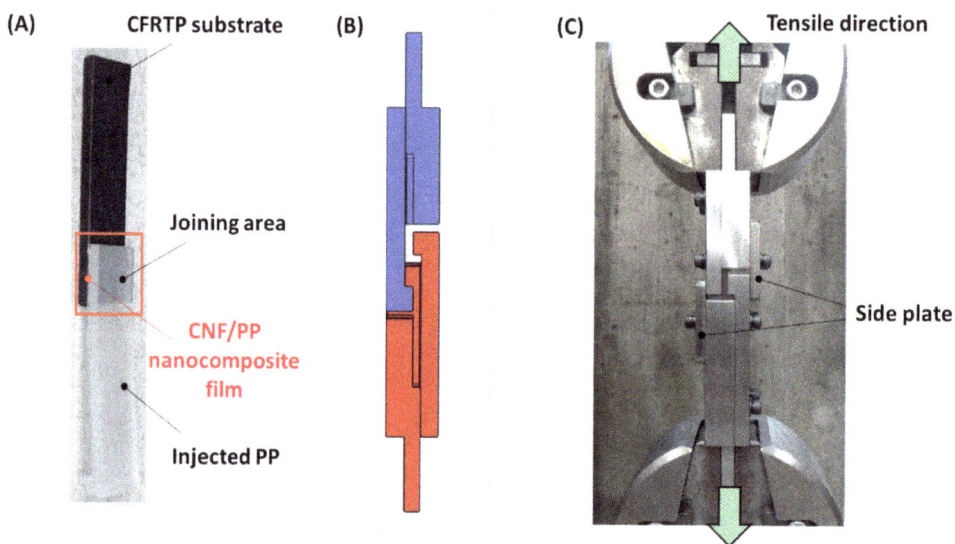

Figure 1: (a) The fabricated single lap joint through the injection overmolding process; (b) the cross-section view of the self-designed fixture for the tensile shear test; and (c) the actual fixture with side plates set in the universal testing machine.

2.3 Tensile shear creep test

The tensile shear creep test was conducted using a tensile creep apparatus (CREEP TESTER L100ER, Toyo Seiki Seisaku-sho Ltd., Japan). Single lap joints fabricated by an injection speed of 80 mm/s and cylinder temperature of 260°C with various CNF contents were used for the creep test. First, to identify the creep rupture life, 60% to 80% load against the maximum load obtained from the tensile shear test was applied at 30°C. The creep test was repeated three times for each specimen. Second, the influence of temperature on creep deformation under a constant load of 625 N (stress of 5.0 MPa). The temperature was varied from 30 to 110°C with an interval of 20°C. The debonded surface was observed through OM (SZX7, Olympus Corporation, Japan) and FE-SEM (S-4000, Hitachi High-Tech Corporation, Japan).

3 RESULTS AND DISCUSSION

3.1 Influence of molding conditions and CNF content on the joining strength

The results of lap shear strength under various injection speeds and cylinder temperatures and various CNF contents in the CNF-filled PP nanocomposite film are summarized in Fig. 2. From the viewpoint of molding conditions, the higher cylinder temperature and higher injection speed increased lap shear strength. Pisanu et al. [16] also reported that higher holding pressure, cylinder temperature and injection speed increased the joining strength.

To determine the pressure and temperature state at the joining interface during injection overmolding, an example of process data at a cylinder temperature of 260°C and an injection speed of 80 mm/s is presented in Fig. 3(a). After the melt polymer is filled in the mold cavity,

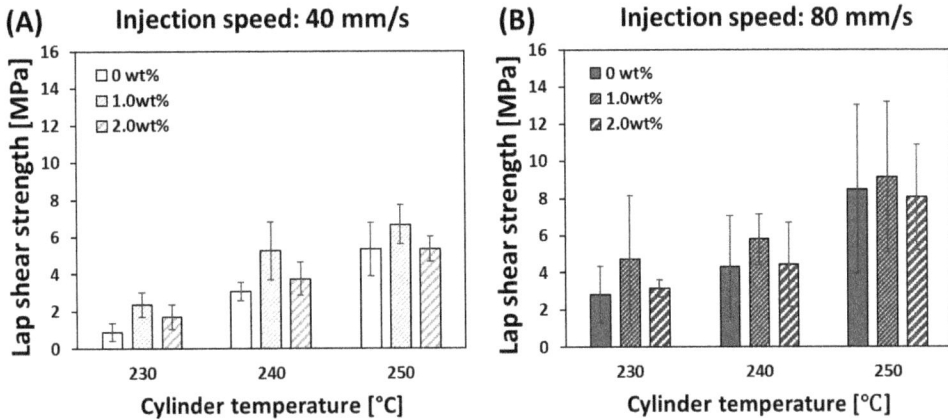

Figure 2:   The lap shear strength of a single lap joint molded under various cylinder temperatures at: (a) an injection speed of 40 mm/s; and (b) an injection speed of 80 mm/s.

the resin temperature and cavity pressure suddenly increase. Following that, the temperature and pressure gradually decreased during the holding stage since the injected PP may be gradually solidified and shrank by cooling of PP in the mold. To achieve a higher joining strength, the development of intimate contact and molecular interdiffusion at the joining interface provides enough pressure and temperature, respectively [9]. Here, the maximum resin temperature and cavity pressure are summarized in Fig. 3(b) and 3(c), respectively. The higher cylinder temperature significantly increased the resin temperature at the joining area, while the cylinder temperature did not influence the cavity pressure. In the influence of the injection speed, the cavity pressure was increased, while the resin temperature was not affected. Thus, a higher cylinder temperature contributed to increasing the resin temperature, and a higher injection speed contributed to increasing the cavity pressure. However, the resin temperature was lower than the cylinder temperature by approximately 10°C. The interface temperature (T_i) is predicted by the following equation [17]:

$$T_i = \left(T_p + T_0\right)/2, \tag{1}$$

where T_p is the resin temperature and T_0 is the mold temperature. Thus, the range of interface temperature was 140–150°C against the range of cylinder temperature of 230–260°C in this study. Since the peak of the melting temperature of PP is assumed to be nearly 160°C, a higher T_i is necessary to achieve a higher joining strength from the viewpoint of molecular interdiffusion.

In the influence of the CNF content, the addition of 1.0 wt% CNF to the PP nanocomposite film increased the joining strength under all process conditions. Furthermore, the addition of 2.0 wt% CNF deteriorated the lap shear strength compared with the 1.0 wt% CNF-filled single lap joint, as reported in our previous work [13], [14]. Thus, these results showed that the optimum content of CNFs existed. However, a large standard deviation was confirmed for all specimens. To achieve more reliable strength and to obtain the reinforcement effect of CNFs, the optimization of molding conditions is supposed to be required for the interface temperature to exceed the melting temperature.

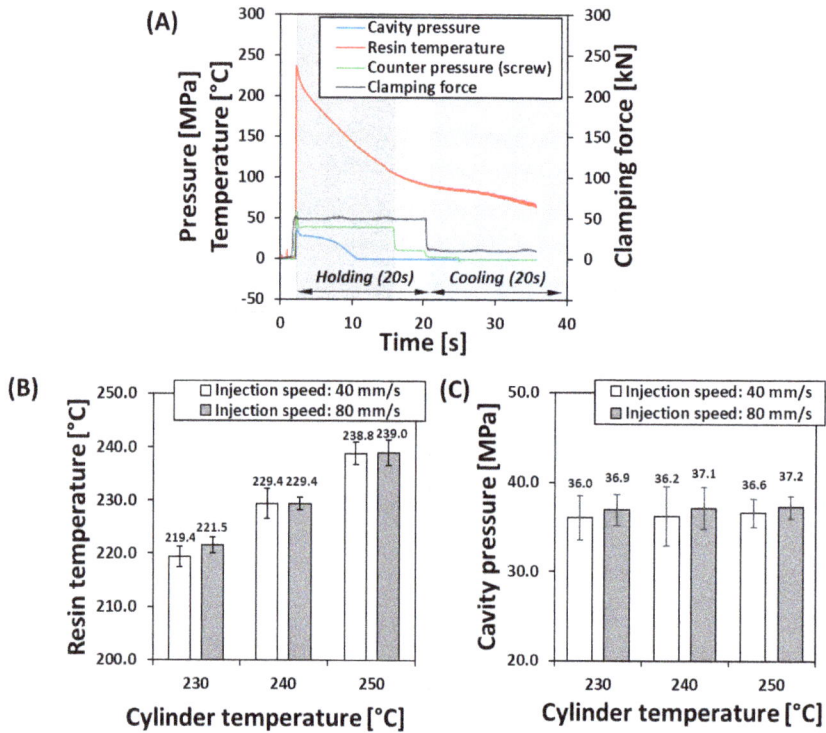

Figure 3: Process characteristics of the injection overmolding process. (a) Process data of one cycle; (b) Maximum resin temperature; and (c) maximum cavity pressure.

3.2 Influence of the applied stress and CNF content on the creep rupture properties

In the creep test, single lap joints molded at a cylinder temperature of 250°C and an injection speed of 80 mm/s were used. To discuss the influence of the CNF content on the creep properties, the samples with the highest lap shear strength among the molding conditions were used. The creep curves of single lap joints with various CNF contents are presented in Fig. 4. The provided stress is also described in the figure. Higher stress decreased the creep rupture time and increased the failure strain for all the samples. However, please note that only the 1.0 wt% CNF-containing sample did not fracture at 60% stress against lap shear strength for 300 h (approximately 12.5 days), and the plot is marked by an arrow (→).

To discuss the influence of the CNF content on the creep rupture time, the creep rupture time was organized by the applied stress, as shown in Fig. 5(a). The relationship between the applied stress $\sigma_{applied}$ and the time to failure in hours t_r was extrapolated by the following power-law equation [18]:

$$\sigma_{applied} = At_r{}^B, \tag{2}$$

where A is the intercept at $t_r = 1$, and B is the slope of the fitted line. The fitted results are also shown in the same figure. The failure time obviously increased at the same stress level with a higher CNF content. Furthermore, the obtained constants against the CNF content are

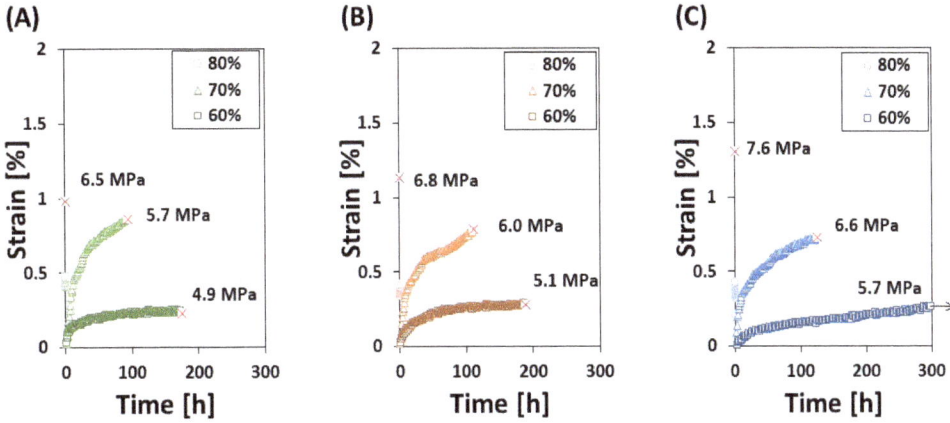

Figure 4:   Creep curves of single lap joints with various CNF contents at the interface nanocomposite film. (a) Without CNF; (b) 0.5 wt%; and (c) 1.0 wt%.

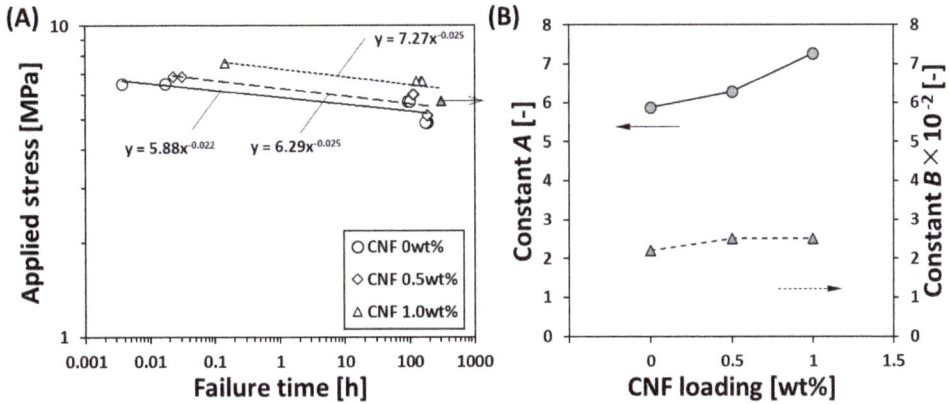

Figure 5:   Creep properties. (a) Relationship between failure time and applied stress; and (b) Constants of extrapolated power law equation.

summarized in Fig. 5(b). The value of A increased with increasing CNF content, while the constant B was slightly influenced by the CNF content. Thus, CNF could contribute to increasing the creep rupture life.

Moreover, the relationship between the applied stress and failure strain is represented in Fig. 6. This result revealed that a higher CNF content decreased the failure strain at the same provided stress. This means that CNFs contributed to suppressing the shear deformation of PP at the joining interface. The debonded surface observed through OM is presented in Fig. 7. Please note that the debonded surface was observed from the CFRTP substrate part. Interestingly, the CNF-containing samples showed many cracks compared with the sample without CNFs. Furthermore, the debonded surface observed through SEM is shown in Fig. 8. The debonded surface and crack region were observed. By the addition of CNFs, the

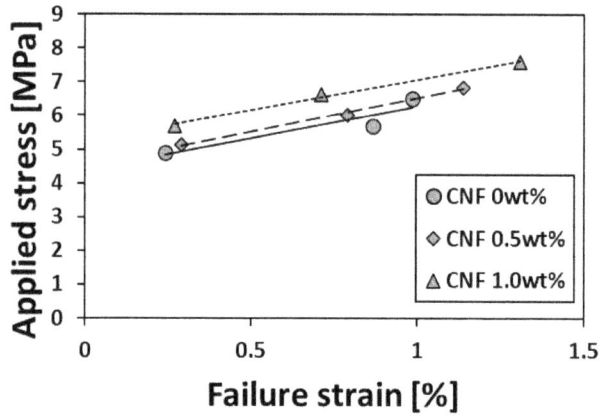

Figure 6: The relationship between the applied stress and failure strain of single lap joints with various CNF contents.

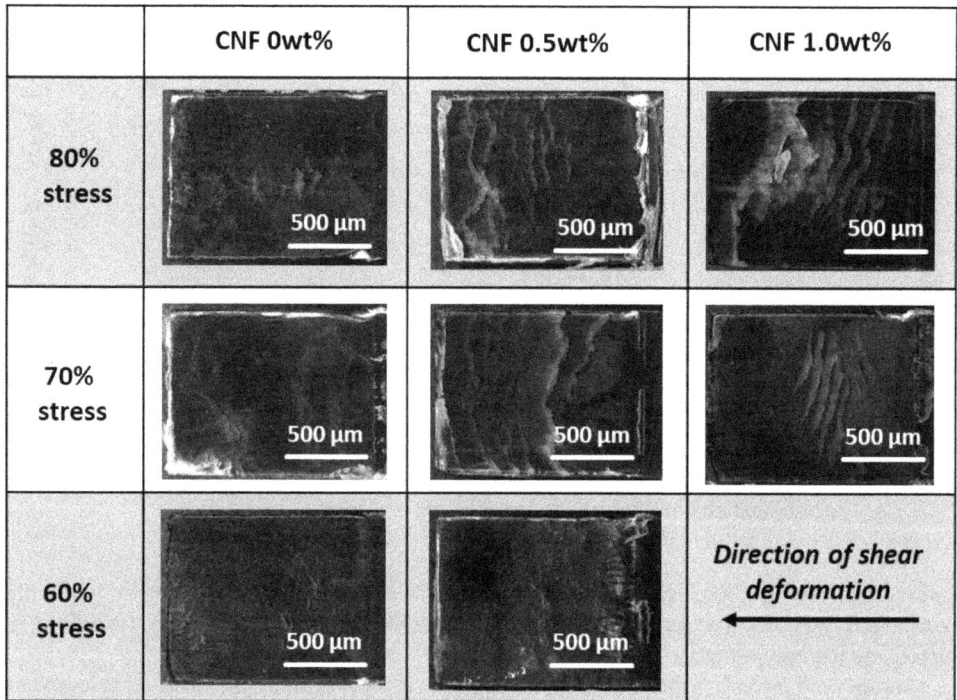

Figure 7: The debonded surface observed through OM from the side of the CFRTP substrate.

debonding surface was changed to a rough surface. Furthermore, fiber bridging was observed upon CNF addition. Thus, this debonded surface may indicate that CNFs contribute to interconnecting the CFRTP substrates and injected polymer or restricting the shear deformation of the polymer at the interface.

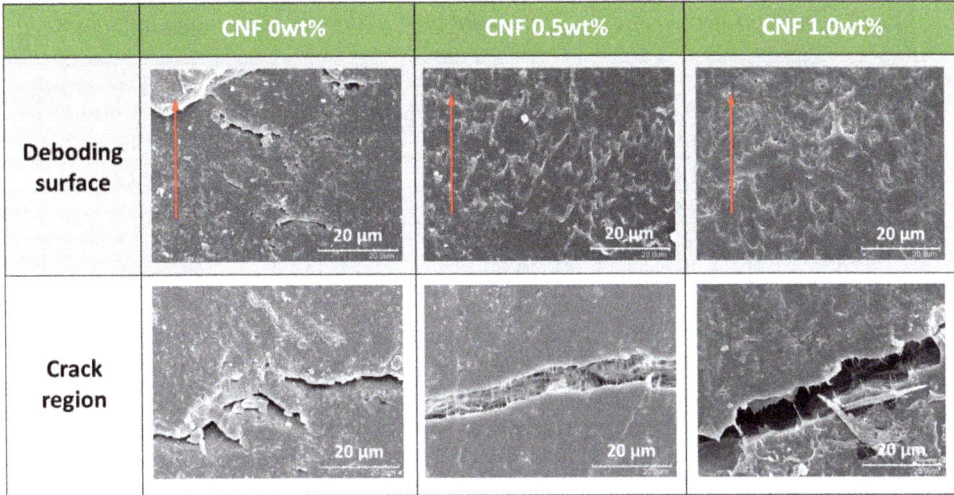

Figure 8:   The debonded surface observed through FE-SEM from the side of the CFRTP substrate. The arrows indicate the direction of shear deformation.

3.3 Influence of temperature and CNF content on the creep properties

The creep deformation curves under various temperatures are presented in Fig. 9. The creep tests were conducted at a constant stress of 5.0 MPa for 48 h, and only the single lap joint without CNF at 110°C was debonded. Regardless of the temperature, CNF addition decreased the creep strain. Furthermore, the initial strain ε_0 and creep strain rate at the secondary creep stage are summarized in Table 2.

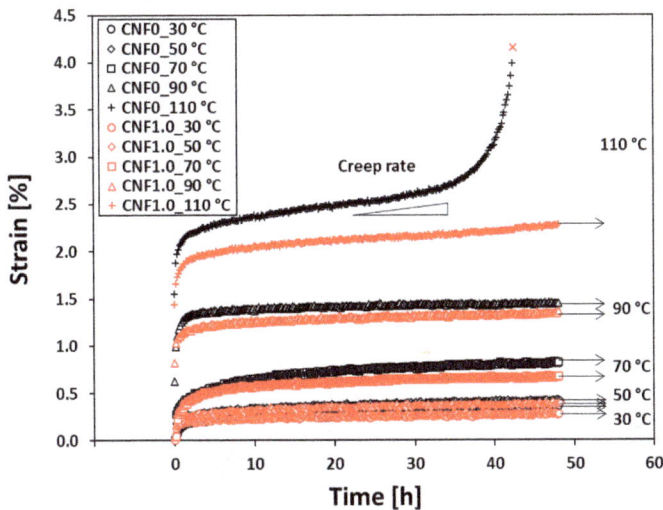

Figure 9:   The creep curves of a single lap joint with 1.0 wt% CNF and without CNF under various temperatures.

Table 2: Creep properties of a single lap joint with CNFs under various temperatures.

Sample	CNF 0 wt%		CNF 1.0 wt%	
Temperature (°C)	ε_0 (%)	Creep rate $\times 10^{-5}$ (1/h)	ε_0 (%)	Creep rate $\times 10^{-5}$ (1/h)
30	0.14	1.97	0.13	0.96
50	0.18	1.72	0.17	0.95
70	0.28	2.63	0.22	1.21
90	0.62	2.07	0.82	1.72
110	1.55	10.99	1.44	5.44

The initial strain and creep rate increased with increasing temperature, as reported elsewhere. While the CNF addition did not influence the initial strain, the creep rate significantly decreased with the addition of 1.0 wt% CNFs regardless of temperature. Thus, we found that the CNFs evidently retarded creep deformation even though the temperature was relatively high. From the rheological point of view, creep deformation under the shear mode occurred at the joining interface. Wang et al. [19] investigated the melt creep properties of CNF/PP nanocomposites, and they reported that 10 wt% CNF decreased the creep strain since the CNFs restricted the movement of the polymer chains. Thus, the restriction of chain movement by CNFs may contribute to decreasing the creep rate.

3.4 The mechanism of property improvement by the addition of CNF

From the actual resin temperature during injection overmolding (Fig. 3(a)), the interface temperature may be less than the melting peak temperature of PP. Thus, the molecular interdiffusion of PP at the joining interface was not perfectly achieved in this study. However, the debonded surface had a rough surface by addition of CNFs, as shown in Fig. 8, even though the debonded surface of the CFRTP substrate without CNFs showed a smooth surface. The rough surface was assumed to be made by local shear deformation of PP from the existing position of individual CNFs or CNF agglomerates.

We assumed the following joining mechanism by using CNFs, as shown in Fig. 10. First, (I) heat transfer occurs during injection overmolding from the injected melt polymer to the CNF-filled PP nanocomposite layer. (II) The PP in the nanocomposite layer was partially melted by heat transfer, and many nano- to microlevel grooves were made close to the individual CNFs and CNF agglomerates. The injected polymer entered the grooves by injection screw pressure at the holding stage, and a larger contacting surface area was made at the joining interface. (III) The large joining surface area and the restriction of PP chain movement by CNF suppress the shear deformation during the tensile shear test. Consequently, the many convex parts at the joining interface were deformed in the shear direction, and cracks are generated. The variation in joining interface topology by the

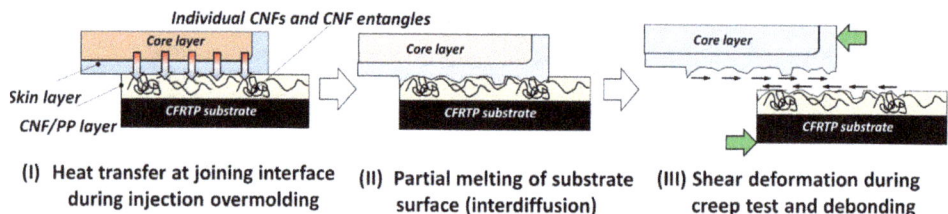

Figure 10: The joining mechanism by addition of CNF in injection overmolding process.

addition of CNFs should be ensured by using a nondestructive method. Furthermore, the crystalline morphology of PP also affects the joining mechanism since the CNFs have a nucleation effect and impact the melting behavior during the welding process. The influence of the process on the morphology of the joining interface, which contained nanofibers, should be clarified.

4 CONCLUSIONS

This study aimed to describe the effects of CNF addition at the joining interface of a single lap joint fabricated by an injection overmolding process on the creep properties. The self-designed fixture was used for the tensile shear test to investigate the lap shear strength and creep properties. Before conducting the creep test, the influence of the injection speed and cylinder temperature on the joining strength was investigated. The higher injection speed and cylinder temperature exhibited higher joining strength since higher pressure and resin temperature were achieved. Regardless of the molding conditions, the addition of 1.0 wt% CNF showed the highest joining strength in this study.

In the shear creep test, the specimens that have the highest lap shear strength among the molding conditions were used. Various constant loads and various temperatures were provided. We found that CNF addition increased the creep failure time and reduced the creep strain against the same stress. Furthermore, the CNF contributed to decreasing the creep rate at the secondary stage regardless of temperature. Furthermore, many cracks and rough surface states were observed at the debonded surface by the addition of CNFs.

However, the interface temperature was lower than the melting peak temperature of used PP in a used molding condition. Thus, to obtain the maximum effect of nanofiber addition, further optimization of molding conditions is necessary since the joining strength was still lower than the shear strength of PP (approximately 22 MPa). The relationship between creep properties and molding condition should be discussed more. In addition, the influence of nanofiber addition on the fatigue properties should be discussed in the future.

ACKNOWLEDGMENTS

A part of this work was supported by JSPS KAKENHI Grant Number JP21K14060. The authors would like to thank Mitsui Chemicals, Inc., Japan for providing materials. We would like to acknowledge the research institute for engineering at Kanagawa University against the use of FE-SEM.

REFERENCES

[1]   Kleemann, S., Fröhlich, T., Türck, E. & Vietor, T., A methodological approach towards multi-material design of automotive components. *Procedia CIRP*, **60**, pp. 68–73, 2017.

[2]   Cui, X., Zhang, H., Wang, S., Zhang, L. & Ko, J., Design of lightweight multi-material automotive bodies using new material performance indices of thin-walled beams for the material selection with crashworthiness consideration. *Materials and Design*, **32**(2), pp. 815–821, 2011.

[3]   Li, C. & Kim, I.Y., Multi-material topology optimization for automotive design problems. *Proceedings of the Institution of Mechanical Engineers, Part D: Journal of Automobile Engineering*, **232**(14), pp. 1950–1969, 2017.

[4]   Temesi, T. & Czigany, T., Integrated structures from dissimilar materials: The future belongs to aluminum–polymer joints. *Advanced Engineering Materials*, **22**(8), 2000007, 2020.

[5]   Krassmann, D. & Moritzer, E., Development of a new joining technology for hybrid joints of sheet metal and continuous fiber-reinforced thermoplastics. *Welding in the World*, **66**, pp. 45–60, 2022.

[6]   Amancio-Filho, S.T. & Lucian-Attila, B. (eds), *Joining of Polymer-Metal Hybrid Structures: Principles and Applications*, John Wiley & Sons: Hoboken, pp. 101–126, 2018.

[7]   Aliyeva, N., Sas, H.S. & Okan, B.S., Recent developments on the overmolding process for the fabrication of thermoset and thermoplastic composites by the integration of nano/micron-scale reinforcements. *Composites Part A: Applied Science and Manufacturing*, **149**, 106525, 2021.

[8]   Holzinger, M., Blase, J., Reinhardt, A. & Kroll, L., New additive manufacturing technology for fibre-reinforced plastics in skeleton structure. *Journal of Reinforced Plastics and Composites*, **37**(20), pp. 1246–1254, 2018.

[9]   Remko, A., Mark, B. & Sebastiaan, W., Analysis of the thermoplastic composite overmolding process: Interface strength. *Frontiers in Materials*, **7**, p. 27, 2020.

[10]   Jiang, B., Fu, L., Zhang, M., Weng, C. & Zhai, Z., Effect of thermal gradient on interfacial behavior of hybrid fiber reinforced polypropylene composites fabricated by injection overmolding technique. *Polymer Composites*, **41**(10), pp. 4064–4073, 2020.

[11]   Dikshit, V., Bhudolia, S.K. & Joshi, S.C., Multiscale polymer composites: A review of the interlaminar fracture toughness improvement. *Fibers*, **5**(4), p. 38, 2017.

[12]   Kang, M.-H., Choi, J.-H. & Kweon, J.-H., Fatigue life evaluation and crack detection of the adhesive joint with carbon nanotubes. *Composite Structures*, **108**, pp. 417–422, 2014.

[13]   Matsumoto, K., Ishikawa, T. & Tanaka, T., A novel joining method by using carbon nanotube-based thermoplastic film for injection over-molding process. *Journal of Reinforced Plastics and Composites*, **38**(13), pp. 616–627, 2019.

[14]   Matsumoto, K., Nagasaka, T., Takemura, K. & Tanaka, T., Influence of nanofiber loading conditions on the joining strength of thermoplastic composites fabricated by injection over-moulding process. *WIT Transactions on The Built Environment*, vol. 196, WIT Press: Southampton and Boston, pp. 113–124, 2020.

[15]   Pisanu, L., Santiago, L.C., Barbosa, J.D.V., Beal, V.E. & Nascimento, M.L.F., Strength shear test for adhesive joints between dissimilar materials obtained by multicomponent injection. *International Journal of Adhesion and Adhesives*, **86**, pp. 22–28, 2018.

[16]   Pisanu, L., Santiago, L.C., Barbosa, J.D.V., Beal, V.E. & Nascimento, M.L.F., Effect of the process parameters on the adhesive strength of dissimilar polymers obtained by multicomponent injection molding. *Polymers*, **13**(7), p. 1039, 2021.

[17]   Jiang, B., Fu, L., Zhang, M., Weng, C. & Zhai, Z., Effect of thermal gradient on interfacial behavior of hybrid fiber reinforced polypropylene composites fabricated by injection overmolding technique. *Polymer Composites*, **41**, pp. 4064–4073, 2020.

[18]   Amjadi, M. & Fatemi, A., Creep behavior and modeling of high-density polyethylene (HDPE). *Polymer Testing*, **94**, 107031, 2021.

[19]   Wang, L., Gardner, D.J. & Bousfield, D.W., Cellulose nanofibril-reinforced polypropylene composites for material extrusion: Rheological properties. *Polymer Engineering and Science*, **58**, pp. 793–801, 2018.

KINETIC STREET FURNITURE WITH ARM-Z

ELA ZAWIDZKA & MACHI ZAWIDZKI
Institute of Fundamental Technological Research, Polish Academy of Sciences, Poland

ABSTRACT
Arm-Z is a concept of a hyper-redundant manipulator based on linearly joined sequence of congruent units. Each unit has only one degree of freedom (1-DOF), namely a twist relative to the previous unit in the sequence. Since each module is identical, Arm-Z has a potential of being economical and robust: the modules can be mass-produced and, in case of failure, easily replaced. However, the control of Arm-Z is nonintuitive and difficult, thus it usually requires application of computational intelligence methods. This paper presents a number of concepts for kinetic street furniture based on Arm-Z: a spiral column of adjustable height, a sun-tracking shade/solar energy harvester, bio-mimicry sculpture, kinetic sprinkler/fountain. The proposed concepts are low-tech in principle. Therefore in each case, the first module in the sequence is fastened to a solid base (ground). For simplicity, the drive is applied directly to the first module and transferred to subsequent units by internal gears. Each module is equipped with a set of cylindrical and bevel gears with straight teeth with involute profile (for connecting the modules).
Keywords: Arm-Z, extremely modular system, low-tech, street furniture.

1 INTRODUCTION
Sophisticated 3D tubular shapes can be built with simple congruent modules, as presented in Fuhs and Stachel [1]. An analogous parametric design system for creating 3D mathematical knots composed of only one type of unit has been introduced in Zawidzki and Nishinari [2]. Arm-Z is a concept of robotic manipulator based on the same idea, which has been introduced in Zawidzki and Nagakura [3]. Biological snakes are extremely well-adapted for various types of environment. It is mostly due to the high redundancy of the snake mechanism. In many cases of irregular environments the bio-inspired robots outperform traditional robots equipped with wheels, legs or tracks. The research on snake-like robots is carried out for several decades. This type of locomotion has been researched already in the 1940s [4]. Fifty years later a rigorous mathematical model of this locomotion has been developed. In the late 1990s, a trunk-like locomotors and manipulators have been introduced in Hirose [5]. Snake-like motion gives this type of manipulators certain advantage over conventional robotic manipulators in various environments. They are capable of operating in geometrically complex environments which are inaccessible for conventional robots and manipulators. Such snake-like robotic manipulators can be equipped with various types of working heads for: surveillance, cleaning, welding, etc., as shown in Fig. 1.

Conventional industrial manipulators have low number of degrees of freedom (DOF). On the other hand, bionic trunk-like or snake-line robotic arms have large (redundant) number of DOF. Arm-Z has as many DOFs as the number of modules minus one. Therefore Arm-Z can be categorized as a so-called *hyper* redundant manipulator (HRM) [6]. The inverse kinematic problem of a typical industrial manipulator can be solved easily [7], therefore, its control is straightforward. Conversely, since HRMs are highly non-linear, their control is not straightforward at all, and requires application of artificial intelligence methods [8]–[10]. For more information on this type of manipulators see Chirikjian and Burdick [11].

2 THE CONCEPT OF ARM-Z
Arm-Z modules are geometrical parametric objects analogous to sectors of a circular torus. Each module is defined by the following parameters: size r, offset d, and angle between

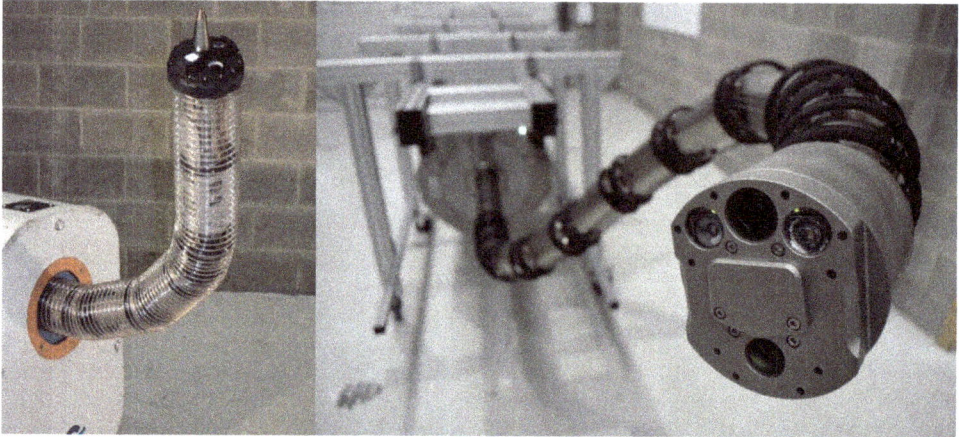

Figure 1: Oliver Crispin Robotics Ltd. Snake-arm robots, series II, X125 System. On the left: sleeved and integrated manipulator with a laser cutting head. On the right: unsleeved, integrated manipulator with inspection camera and light tool. *(Source: http://www.ocrobotics.com.)*

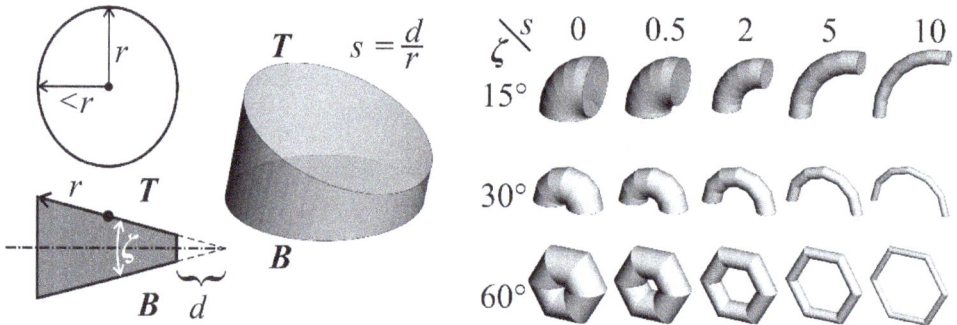

Figure 2: The geometric parameters defining Arm-Z module. On the right: a table with simple sequences of modules with various angles ζ and slenderness s.

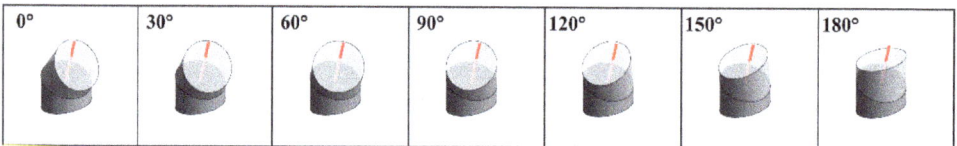

Figure 3: The bottom (dark gray) module is fixed to the base and does not rotate. The axis of rotation, which is perpendicular to the bottom face of the top module is shown in red.

bottom (**B**) and top (**T**) faces of the module – ζ, as illustrated in Fig. 2. Slenderness (s) is an additional parameter, that is a d to r ratio.

The overall shape of Arm-Z is a function of: the number of modules, the geometric parameters the module, and the sequence of relative twists between each pair of modules. Fig. 3 shows two modules at six successive twists from 0 to 180° in 30° steps.

3 STREET FURNITURE WITH ARM-Z

This section presents a few concepts of low-tech street furniture based on Arm-Z.

3.1 A spiral column of adjustable height

Columns are elements which transmit to structural members below the weight of the structure above. This architectural invention allows supporting of ceilings without the use of solid

walls. Therefore the space spanned by a ceiling can increase. The first use of columns was as a single central support for the roof of relatively small buildings. More elaborated columns with aesthetic function, that is beyond mere structural support emerged since the Bronze Age (3000–1000 BC) in Minos, Assyria and Egypt.

The Solomonic column is a helical column in a form of a spiraling twisting corkscrew-like shaft. Probably the most famous Solomonic columns are Bernini's colossal bronze composite columns of the Baldachin in Saint Peter's Basilica (see Fig. 4).

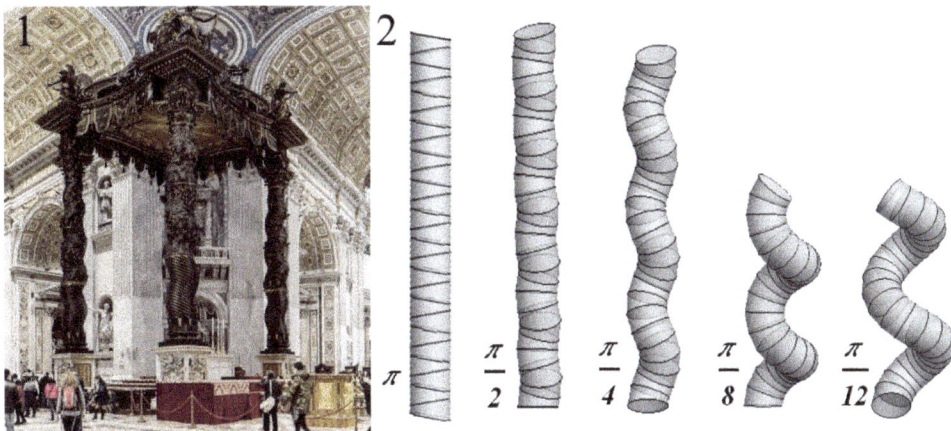

Figure 4:   1. Historical example of spiral columns: Bernini's Baldachin in St. Peter's Basilica (AD 1634). 2. Five examples of Arm-Z kinetic spiral columns controlled by relative twists (same for each unit, the values are shown for each case).

3.2 A sun-tracking shade/solar energy harvester

The prototype of the Arm-Z modular solar tracker has been described in our previous paper [12]. The purpose of such a device is to efficiently harvest solar energy or to serve as an active sun-shade. Fig. 5 on the left shows the sun direction to be followed by Arm-Z during summer (from the solstice to autumn equinox) between 10:00 and 18:00 hours. The positioning of Arm-Z in space and illustration of the tracking parameters are shown on the right.

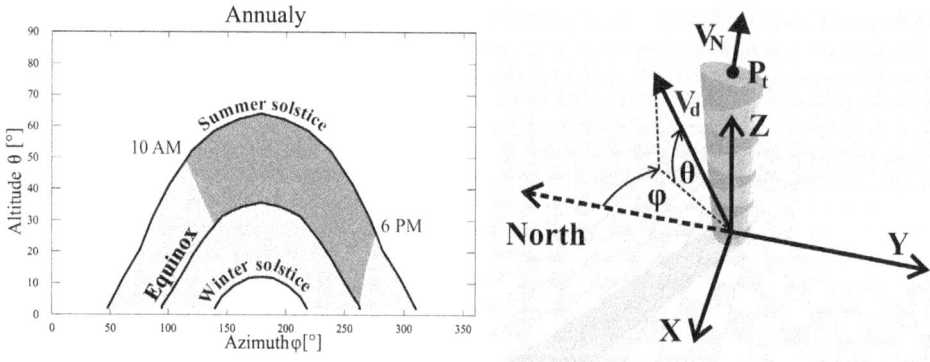

Figure 5:   On the left: the sun positions during a year. Gray hatch indicates the considered tracking periods. On the right: the coordinate system, nomenclature and controlling parameters of the Arm-Z. P_t is the position of the tip of Arm-Z. V_N is the vector directing from the tip of Arm-Z. V_d is the sunlight direction. V_d (φ, θ) is a function of the azimuth angle φ and altitude θ.

The sun tracking is formulated here as minimization of the angle between vectors V_N and V_d. This optimization problem has been solved in Zawidzka et al. [12] using dual annealing [13], [14]. Authors show there that Arm-Z with only four modules (three twisting and one fixed to the base) is capable of positioning its tip in almost all required directions. Fig. 6 shows more results of this optimization.

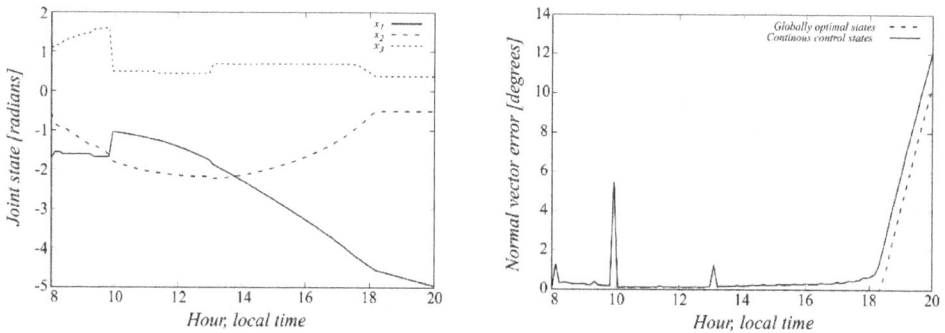

Figure 6:   On the left: the relative rotations of the three twisting modules: x1, x2, x3 for a 4-unit Arm-Z following the sun on 1 July between 08:00 and 20:00. On the right: the errors, that is the differences between vector V_N and desired direction V_d for globally optimal states and for "smooth" action scenario.

A low-tech prototype of Arm-Z solar tracker has been designed and presented in Zawidzka et al. [12]. It was equipped with gears allowing for switching the spin of each unit (left/right) about its axis and transferring of the rotation to the next unit. The gear train has been equipped with so called "reverse switch". A special right/left lever switch has been placed in the side of each unit (see Fig. 7).

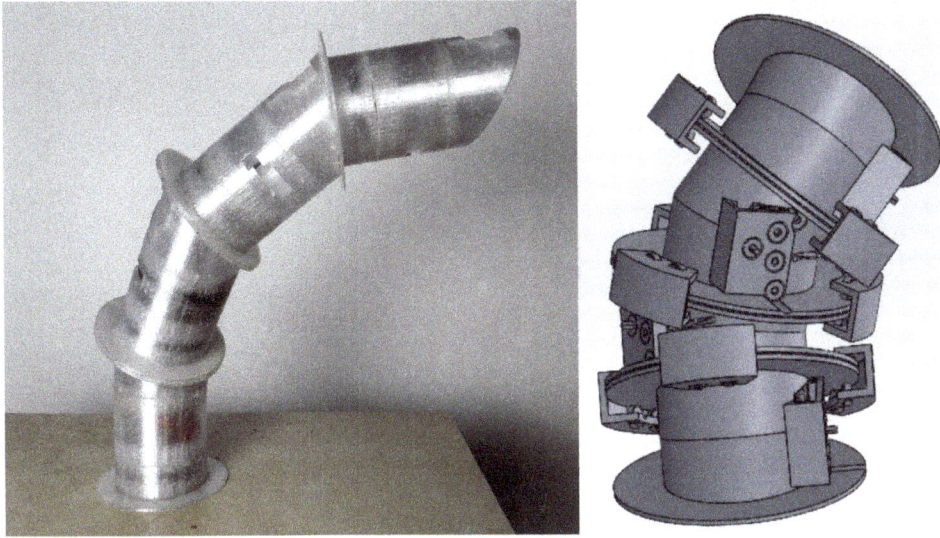

Figure 7:   On the left: a photograph of preliminary 4-module Arm-Z. On the right: a computer model of a 4-module Arm-Z solar tracker.

It is would also be reasonable to install a larger Sun energy harvesting and/or shading element on the tip of Arm-Z, as shown in Fig. 8. In case of larger such elements it is also conceivable to synchronize three or more Arm-Zs as shown in the same figure.

Figure 8:   On the left: a single Arm-Z with larger shading/PV element. On the right (an elastic?) canopy stretched between three synchronized Arm-Zs.

3.3 Bio-mimicry sculpture

An Arm-Z with a dozen or so modules can perform a bio-mimic underwater seed-like motion. Fig. 9 shows selected time-steps of a random motion of a 12-module Arm-Z.

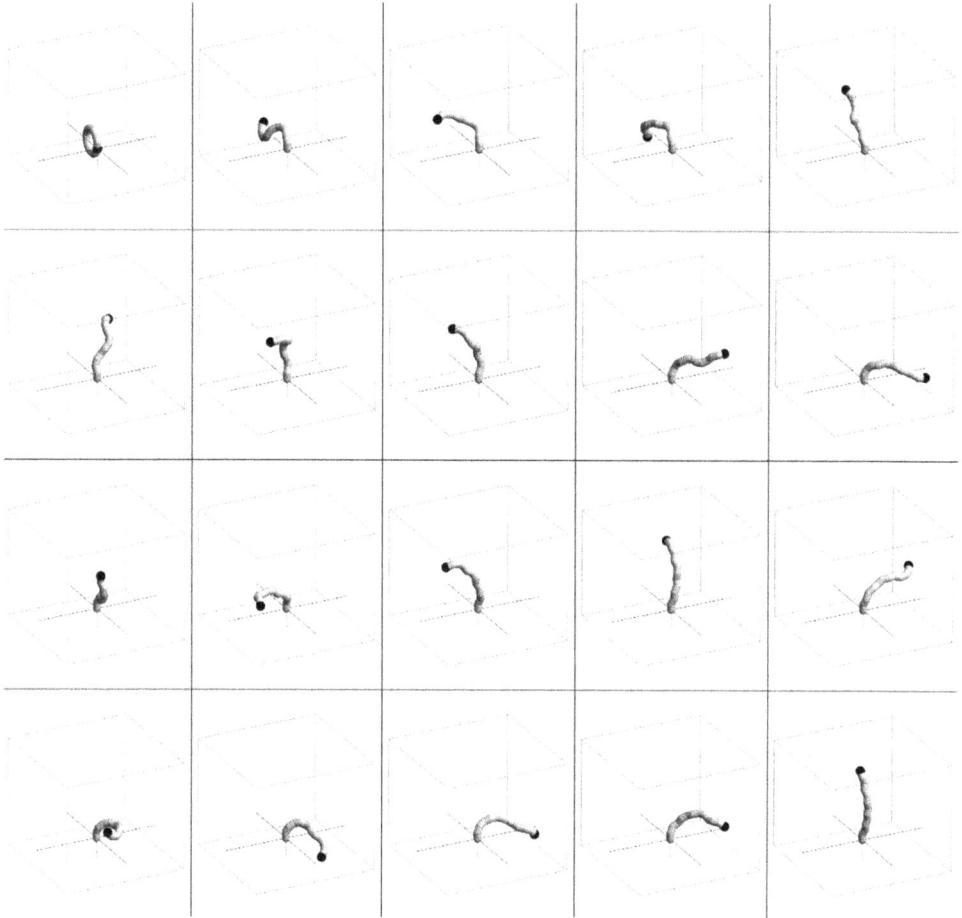

Figure 9: A 12-module Arm-Z "worm" in a random movement.

For an interactive demonstration of this Arm-Z see Zawidzki [15]. However, this concept would require either interdependent control of each module, or random (to a certain degree) rotation of each module. In any case it seems like a major challenge at this point.

3.4 Kinetic Arm-Z sprinkler/fountain

One of the approaches in building the arm-Z prototype was to place the gear system close to the external casing and leaving the center of the module hollow as shown in Fig. 10.

Such a sprinkler-Arm-Z could work either as a simple watering device performing a relatively straightforward circular motion, or could perform of an "unwinding/winding" motion, as illustrated in Fig. 11.

This is particularly interesting case, as it is relatively straightforward to make. From a torus configuration, where initially each subsequent module is at 0° of relative angle to the previous one, all modules simply perform a simultaneous 360° relative rotation and the entire structure returns to the initial toric state.

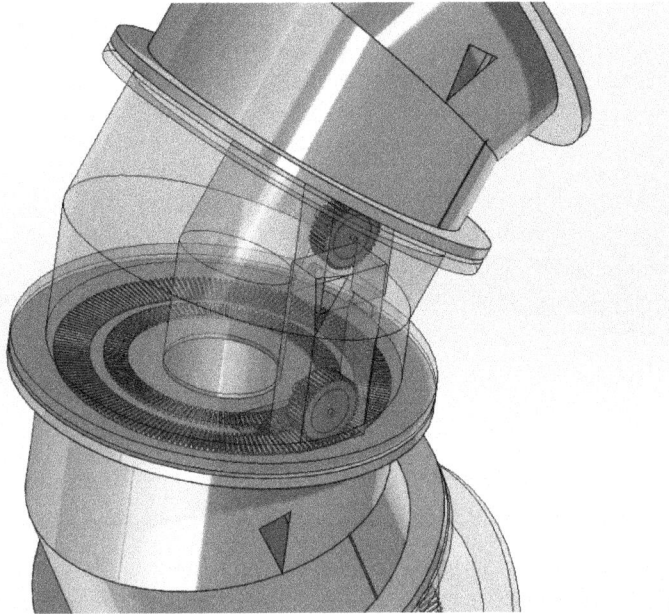

Figure 10: Hollow space inside each module of this type of Arm-Z module allows for installing a water-pipe.

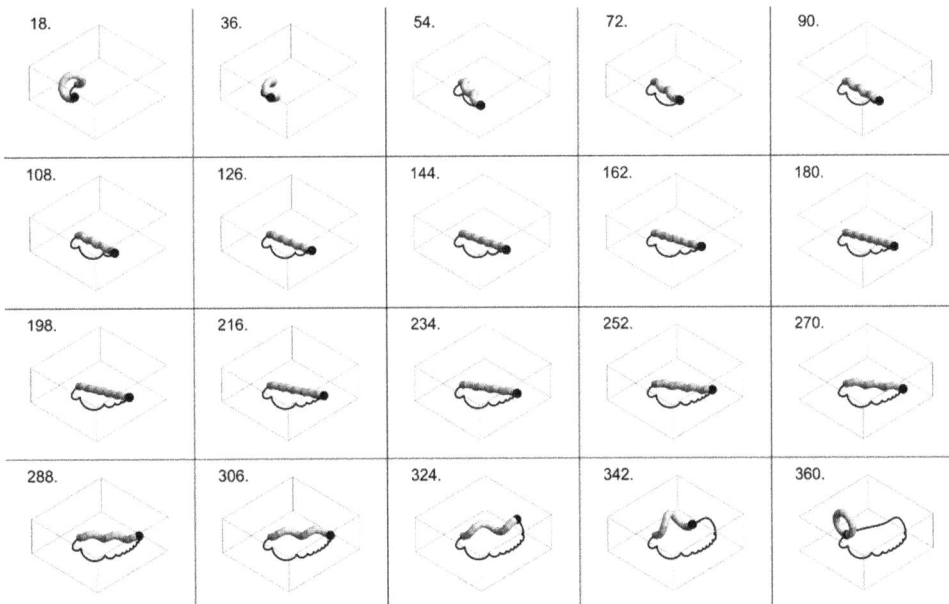

Figure 11: Selected time-steps of "unwinding" of a 12-module Arm-Z from torus into a straight pipe and back to torus. The relative twist is the same for each module and the value in degrees is shown for each frame. The trace of the tip of Arm-Z is shown as a black line.

4 CONCLUSIONS

- Arm-Z is in principle a very simple system, however, it can produce interesting behavior.
- These properties can be used for low-tech street furniture.
- Four types of such architectural elements have been presented.
- The prototypes are presently being developed.

ACKNOWLEDGEMENTS

This research is a part of the project titled "Arm-Z: an extremely modular hyper-redundant low-cost manipulator – development of control methods and efficiency analysis" and funded by OPUS 17 research grant No. 2019/33/B/ST8/02791 supported by the National Science Centre, Poland.

REFERENCES

[1] Fuhs, W. & Stachel, H., Circular pipe-connections. *Computers and Graphics*, **12**(1), pp. 53–57, 1988.

[2] Zawidzki, M. & Nishinari, K., Modular pipe-z system for three-dimensional knots. *Journal for Geometry and Graphics*, **17**(1), pp. 81–87, 2013.

[3] Zawidzki, M. & Nagakura, T., Arm-Z: A modular virtual manipulative. *Proceedings of the 16th International Conference on Geometry and Graphics*, pp. 75–80, 2014.

[4] Gray, J., The mechanism of locomotion in snakes. *Journal of Experimental Biology*, **23**(2), pp. 101–120, 1946.

[5] Hirose, S., *Biologically Inspired Robots: Snake-Like Locomotors and Manipulators*, Oxford University Press, 1993.

[6] Ning, K. & Worgotter, F., A novel concept for building a hyper-redundant chain robot. *IEEE Transactions on Robotics*, **25**(6), pp. 1237–1248, 2009.

[7] Murray, R.M., Li, Z., Shankar Sastry, S. & Shankara Sastry, S., *A Mathematical Introduction to Robotic Manipulation*, CRC Press, 1994.

[8] Rolf, M. & Steil, J.J., Efficient exploratory learning of inverse kinematics on a bionic elephant trunk. *IEEE Transactions on Neural Networks and Learning Systems*, **25**(6), pp. 1147–1160, 2014.

[9] Melingui, et al., Qualitative approach for forward kinematic modeling of a compact bionic handling assistant trunk. *IFAC Proc. Volumes*, **47**(3), pp. 9353–9358, 2014.

[10] Falkenhahn, V., Hildebrandt, A., Neumann, R. & Sawodny, O., Dynamic control of the bionic handling assistant. *IEEE/ASME Transactions on Mechatronics*, **22**(1), pp. 6–17, 2017.

[11] Chirikjian, G.S. & Burdick, J.W., A hyper-redundant manipulator. *IEEE Robotics and Automation Magazine*, **1**(4), pp. 22–29, 1994.

[12] Zawidzka, E., Szklarski, J., Kiński, W. & Zawidzki, M., Prototype of the Arm-Z modular solar tracker. *Conference on Automation*, Springer: Cham, pp. 273–282, 2022.

[13] Xiang, Y., Sun, D.Y., Fan, W. & Gong, X.G., Generalized simulated annealing algorithm and its application to the Thomson model. *Physics Letters A*, **233**(3), pp. 216–220, 1997.

[14] Virtanen, et al., SciPy 1.0: Fundamental algorithms for scientific computing in Python. *Nature Methods*, **17**, pp. 261–272, 2020.

[15] Zawidzki, M., Modeling of Arm-Z with AnglePath3D. Wolfram Demonstrations Project, 2020. http://demonstrations.wolfram.com/ModelingOfArmZWithAnglePath 3D/.

INFLUENCE OF SUPERPLASTICIZER TYPE AND DOSAGE ON RETENTION OF CONSISTENCY OF RUBBERIZED CONCRETE

IVANA MILIČEVIĆ[1], ROBERT BUŠIĆ[1], KRISTIJAN BEBEK[2] & DAVID BRIŠEVAC[1]
[1]Faculty of Civil Engineering and Architecture Osijek, Josip Juraj Strossmayer University of Osijek, Croatia
[2]Bt3 Betontechnik GmbH, Austria

ABSTRACT

Concrete with recycled rubber as a partial replacement of fine natural aggregate, intended for use in load-bearing structural elements, requires specific fresh concrete properties such as a particular slump and flow classes. These specific requirements can be caused by increased water or superplasticizer content. Therefore, concrete mixtures with different percentages of two types of superplasticizers A1 and A2, i.e. 0.2, 0.4 and 0.6% by cement mass, were prepared. Furthermore, the effect of retention of consistency after 15, 30, 45, 60, and 90 minutes, was also studied. Slump and flow table tests were performed at 15-minute intervals to determine the fresh performance of each concrete mixture and retention of consistency. Test results indicate that mixtures with superplasticizer A2 show a more uniform workability range during the measurement period and at the end of the measurement, remain within the same consistency classes, as in the initial measurement.
Keywords: recycled rubber aggregate, superplasticizer, slump, workability, retention of consistency.

1 INTRODUCTION

Due to a large number of unusable waste tires from different types of vehicles, common disposal methods such as landfilling and incineration can cause serious ecological problems, either because of rapid site depletion or air pollution [1]–[3]. Various associations around the world are promoting a circular economy and sustainable development, recycling of scrap tires, and reuse of tires by adding new value to the recycled material. One of the ways to reuse waste rubber is to add it to concrete as a replacement for natural aggregates. The use of recycled rubber as a partial aggregate replacement has the following effects on concrete: reduction of water content, reduction of workability, and may affect mechanical properties due to insufficient bonds between cement paste and rubber [4]–[7]. Concrete with recycled rubber as a partial replacement for fine natural aggregate intended for use in load-bearing structural elements requires specific fresh concrete properties such as a particular slump and flow class. These specific requirements can be caused by increased water or superplasticizer content. Superplasticizer is a type of water reducer. Superplasticizer significantly reduces the amount of water required when mixing the concrete [8]. The effects of superplasticizer are obvious, that is, producing concrete with very high workability or concrete with very high strength. The mechanism of a superplasticizer is through giving the cement particles a highly negative charge so that they repel each other due to the same electrostatic charge. By deflocculating the cement particles, more water is provided for concrete mixing [9]. Although numerous studies have been conducted with different types of chemical admixtures, very few of them addressed the effect of chemical admixtures on the properties of prolonged mixed concrete [10]–[16]. The objective of this study was to determine the effects of two commercially available superplasticizers on rubberized concrete, focusing on how dosage and type affect the properties of fresh rubberized concrete and retention of concrete consistency during the 90 min period.

WIT Transactions on The Built Environment, Vol 209, © 2022 WIT Press
www.witpress.com, ISSN 1743-3509 (on-line)
doi:10.2495/HPSU220031

2 MATERIAL CHARACTERISTICS AND PREPARATION

2.1 Materials

Concrete mixtures were prepared using recycled rubber 0–4 mm as a partial replacement for fine natural aggregate. Normal Portland cement CEM I 42.5R (characteristic values of mechanical, physical, technical properties, and requirements according to the standard HRN EN 197-1:2011), 5% silica fume (RW-Fuller, properties and requirements according to the standard HRN EN 13263-1:2009), crushed dolomite as coarse aggregate, natural sand from the river Drava as fine aggregate. Two types of the mix were prepared, all with a w/c ratio of 0.45. Mixtures R10-0.2-A1 to R10-0.6-A1 are mixtures corresponding to concrete class C30/37, 160–210 mm slump, 490–550 mm and 1.5% air content, with 10% of crumb rubber and different percentages of superplasticizer A1 (0.2% to 0.6%). Mixtures R10-0.2A2 to R10-0.6-A2 are mixtures in the same ratio as before, but with different percentages of superplasticizer A2 (0.2% to 0.6%). Two different types of superplasticizers were used. Both superplasticizers are based on modified polycarboxylate ethers, which have a strong plasticizing effect of homogenizing concrete. The differences between superplasticizer Energy FM 500 marked as A, and Energy FM 500NX marked as A2 are listed in Table 1.

Table 1: Differences between superplasticizers A1 and A2.

A1	Influence on rheological properties				
	Maintaining level of consistency				
	Early strength				
A2	Influence on rheological properties				
	Maintaining level of consistency				
	Early strength				

2.2 Mixture design and testing method

Table 2 shows the mixture proportions in this experiment. The influence of two types of polycarboxylate superplasticizers A1 and A2 on workability, i.e. slump and flow, was tested.

Table 2: Mixture proportions.

Mark	Weight per unit volume (kg/m^3)							Admixture (%)	
	C	SF	FA	FR	CA	S	W*	A1	A2
R10-CEM-I	427.5	22.5	437.5	65.8	812.9	323.6	212.6	0	0
R10-CEM-I-0.2-A1	427.5	22.5	437.0	65.8	811.9	323.3	212.6	0.2	0
R10-CEM-I-0.4-A1	427.5	22.5	436.4	65.8	810.9	322.8	212.6	0.4	0
R10-CEM-I-0.6-A1	427.5	22.5	435.9	65.8	809.8	322.4	212.6	0.6	0
R10-CEM-I-0.2-A2	427.5	22.5	437.0	65.8	811.9	323.3	212.6	0	0.2
R10-CEM-I-0.4-A2	427.5	22.5	436.4	65.8	810.9	322.8	212.6	0	0.4
R10-CEM-I-0.6-A2	427.5	22.5	435.9	65.8	809.8	322.4	212.6	0	0.6

* k-concept, (w/c = w/(c+k×a))

The ambient temperature was 25°C ± 2°C when the mixtures were prepared. The consistency of the fresh concrete was determined with slump-test in accordance with HRN

EN 12350-2:2019 and flow of fresh concrete in accordance with flow table test HRN EN 12350-5:2019. A total of seven concrete mixtures, with 10% crumb rubber as a partial replacement for fine aggregate and with different admixtures (A1 and A2), were subjected to a mixing period during 90 minutes. Slump and flow tests were carried out for concrete mixtures subjected to prolonged mixing at 15-minute intervals. The concrete was mixed in a pan mixer with a capacity of 120 litres. Each mixture was initially mixed for 8 minutes to ensure its homogeneity. The mixer was stopped at 15-minute intervals to conduct slump and flow tests. The mixing process of each concrete mixture was continued for 90 minutes.

3 RESULTS OF EXPERIMENTAL TESTING

3.1 Influence of superplasticizer type and dosage on concrete workability

Fig. 1 shows examples of slump-tests after 0 minutes, 15 minutes, 30 minutes, 45 minutes, and 60 minutes. It can be seen how the workability of the mixtures decreases with time. As the amount of A1 and A2 adsorbed on the cement particles continues to increase, the slump loss of paste fluidity is slowed. Figs 2 and 3 show the test results of slump-tests of the rubberized concrete mixtures. The data are recorded and shown to observe the relationship between superplasticizer dosage and slump loss.

Figure 1:   Examples of measured slump-test on mixtures. (a) R10-CEM-I; (b) R10-CEM-I-0.2-A1; and (c) R10-CEM-I-0.2-A2.

	0 min	15 min	30 min	45 min	60 min	90 min
■R10-CEM-I	215	190	185	165	155	145
■R10-CEM-I-0.2-A1	200	180	165	155	140	110
□R10-CEM-I-0.4-A1	265	250	245	240	225	210
□R10-CEM-I-0.6-A1	275	270	265	260	255	235

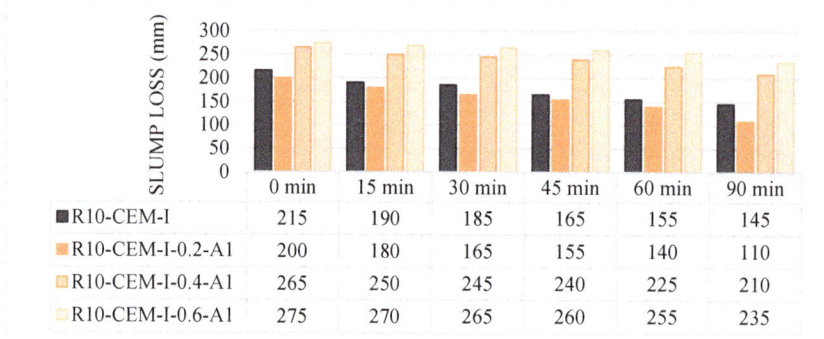

Figure 2: Results of the slump-test for mixtures with A1.

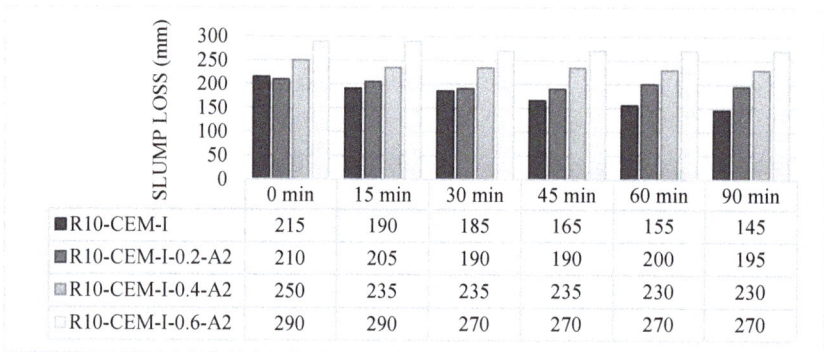

	0 min	15 min	30 min	45 min	60 min	90 min
■R10-CEM-I	215	190	185	165	155	145
■R10-CEM-I-0.2-A2	210	205	190	190	200	195
■R10-CEM-I-0.4-A2	250	235	235	235	230	230
□R10-CEM-I-0.6-A2	290	290	270	270	270	270

Figure 3: Results of the slump-test for mixtures with A2.

Fig. 4 shows the plot between slump loss and superplasticizer dosage at a w/c ratio of 0.45. It can be seen that the minimum slump loss occurs at a dosage of 0.2% and that the optimum superplasticizer dosage for the workability criteria after 90 minutes is 0.4% for A1 and 0.2% for A2 superplasticizer. Fig. 4 shows the change in slump over time for different dosages of superplasticizer. From Fig. 4 it can be seen that the slump decreases with time. This behaviour is acceptable because the continuous hydration process produces calcium silicate hydrate, which fills the pores between the cement particles and the aggregate. As a result, the setting of the concrete will reduce the fluidity of concrete and, consequently, the slump.

Since superplasticizer helps to retain the concrete in liquid state for a longer time, a higher admixture dosage decelerates the setting rate of the concrete. This potentially reduces slump loss during transport of the concrete to the construction site. An overdose of the admixture (0.6% of A1 and A2) leads to a high slump loss, which does not lead to the expected and desired behaviour. On the other hand, it can be concluded that superplasticizer A2 is more effective in maintaining the slump of the concrete with the recycled rubber than superplasticizer A1 or the reference concrete without admixture.

The workability of each mixture was also determined with the flow table. Flow table test was performed according to HRN EN 12350-5:2019. Figs 5 and 6 show the test results of the

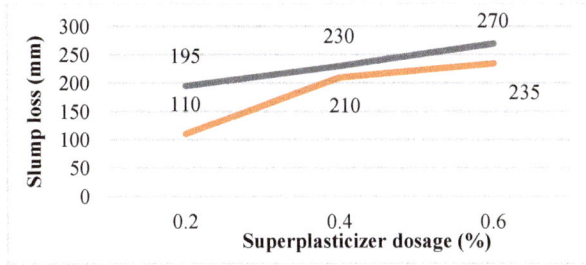

Figure 4: Slump loss vs dosage of superplasticizer.

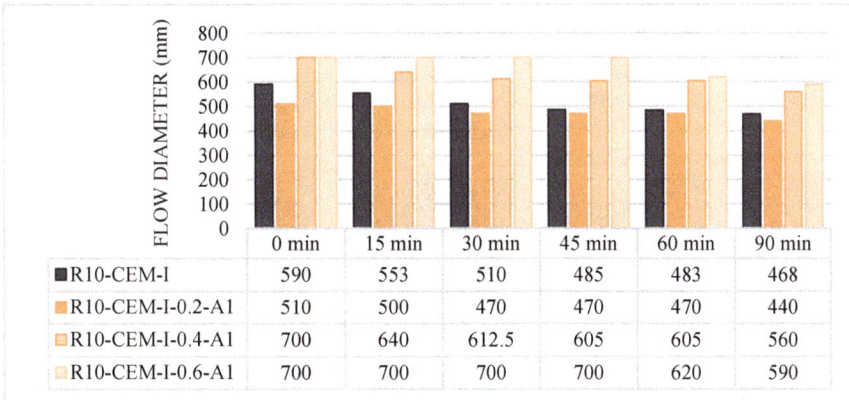

	0 min	15 min	30 min	45 min	60 min	90 min
R10-CEM-I	590	553	510	485	483	468
R10-CEM-I-0.2-A1	510	500	470	470	470	440
R10-CEM-I-0.4-A1	700	640	612.5	605	605	560
R10-CEM-I-0.6-A1	700	700	700	700	620	590

Figure 5: Results of the flow-test for mixtures with A1.

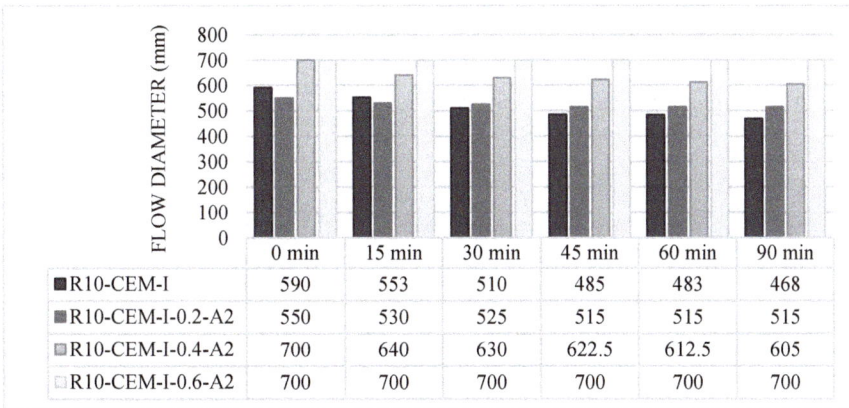

	0 min	15 min	30 min	45 min	60 min	90 min
R10-CEM-I	590	553	510	485	483	468
R10-CEM-I-0.2-A2	550	530	525	515	515	515
R10-CEM-I-0.4-A2	700	640	630	622.5	612.5	605
R10-CEM-I-0.6-A2	700	700	700	700	700	700

Figure 6: Results of the flow-test for mixtures with A2.

tested mixtures. The average of the two measured values of the concrete diameter was recorded, as shown in Figs 5 and 6. There is no visible bleeding or segregation of concrete mixtures, and the use of stabilizing agents was not required. Compared to superplasticizer

A2, a higher dosage of superplasticizer A1 was required to make the paste flow. Concrete mixes with 0.6% A1 and A2 admixtures flow continuously when the cone is removed and topples or shear off before the measurement can be made.

Fig. 7 shows the examples of the measured flow table tests after 0 min, 15 min, 30 min, 45 min and 60 min. The test results shown in Figs 5 to 8 indicate that the flowability of the rubber concrete mix containing superplasticizer A2 is better than that of the concrete containing superplasticizer A1. Superplasticizer A2 keeps the flowability of rubber concrete constant over time, and superplasticizer A1 reduces flowability over time. The presence of superplasticizer A2 with a dosage 0.2% by weight of the cement leads to an increase in the flowability of the concrete mixture compared to that without superplasticizer A2.

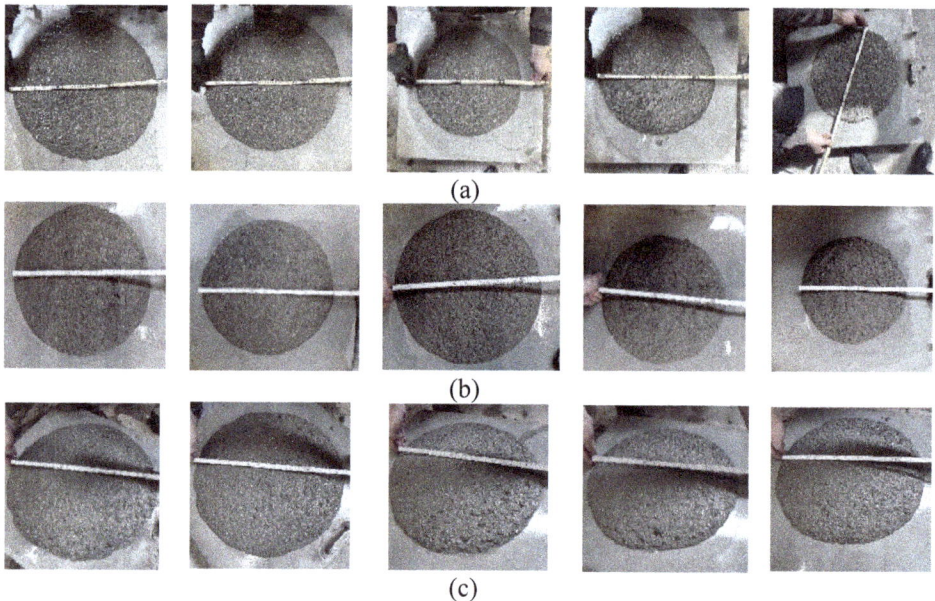

Figure 7: Examples of measured flow table test on mixtures. (a) R10-CEM-I; (b) R10-CEM-I-0.2-A1; and (c) R10-CEM-I-0.2-A2.

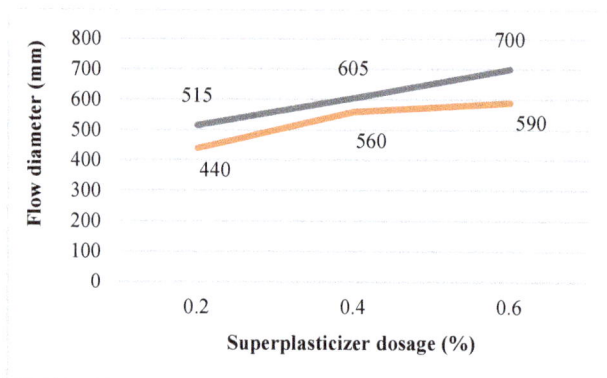

Figure 8: Flow vs dosage of superplasticizer.

3.2 Influence of type and dosage of superplasticizer on retention of concrete consistency

Figs 9 and 10 show the influence of different types and dosages of superplasticizers on retention of the consistency by measuring the slump and flow of each concrete, respectively. The slump behaviour and flowability of concrete mixtures with recycled rubber and superplasticizer A1 are quite changed in the measured periods of the concrete. Mixtures with superplasticizer A2 show a much smaller difference in the change of slump and flow value within the measured time. Moreover, during the measurement period and at the end of the measurement (after 90 minutes), they show a more uniform workability range, within the same consistency classes expressed by slump and flow diameter, as in the initial measurement.

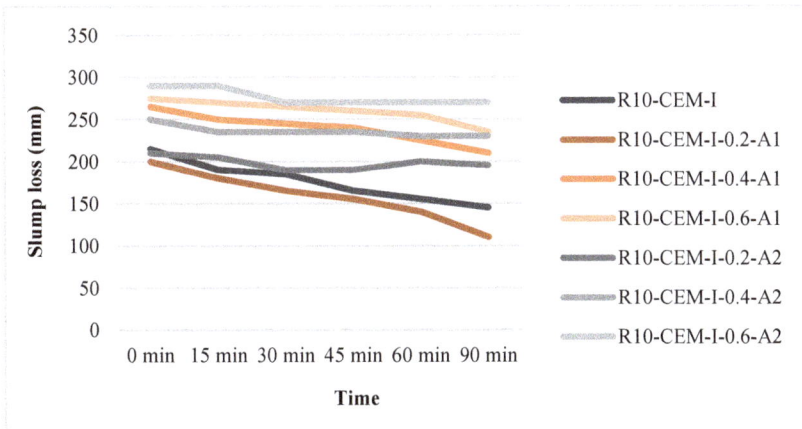

Figure 9: Retention of consistency measured by slump test.

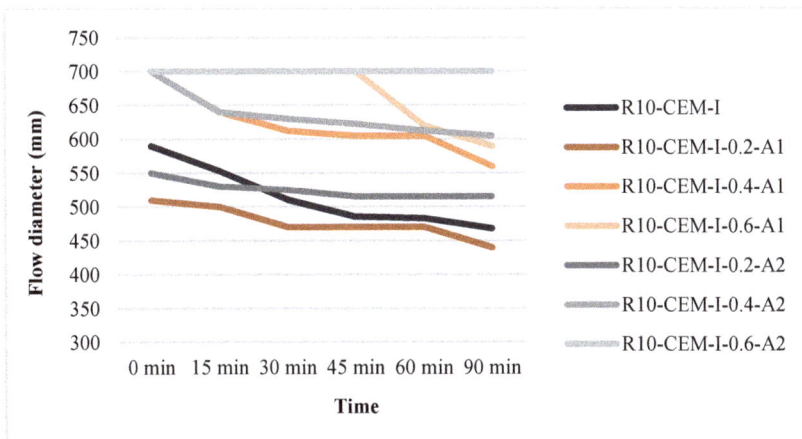

Figure 10: Retention of consistency measured by flow table test.

3.3 Compressive strength

Fig. 11 shows the plot between the strength and dosage of superplasticizer at a 0.45 water–cement ratio. It is evident that the compressive strength of concrete is maximum at 0.4% of superplasticizer dosage and it is obtained as 28.11 and 27.72 MPa. Thus, it can be concluded that there is no significant difference in the compressive strengths between mixtures with superplasticizer A1 and A2.

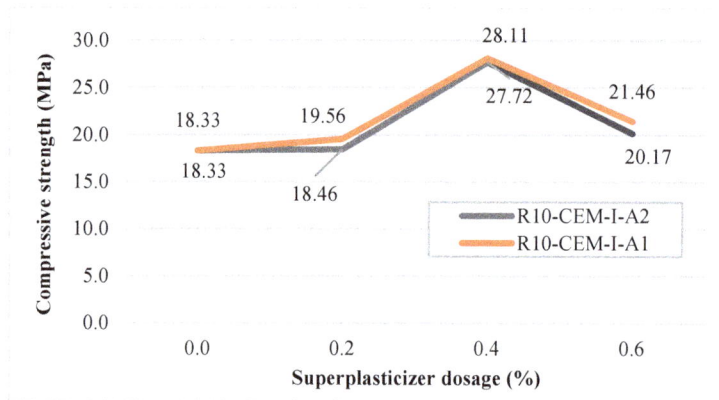

Figure 11: Compressive strength vs superplasticizer dosage.

4 CONCLUSION

With the usage of crumb rubber in concrete, a reduction in workability occurs which can be increased by adding a superplasticizer. However, very high superplasticizer dosages (more than 0.4%) affect the cohesion of the concrete. Slump loss can be reduced by using A1 and A2 superplasticizers. Slump loss and flow diameter test results have shown that the A2 superplasticizer used in this experiment allows concrete containing recycled rubber to be transported over a distance in the range of 90 minutes without the need for delayed or interrupted addition of admixtures. Mixtures with superplasticizer A2 should be highlighted because they show a more uniform workability range during the measurement period and at the end of the measurement (after 90 minutes), remain within the same consistency classes, expressed by slump and flow diameter, as in the initial measurement. Compressive strength is improved by superplasticizer, and there is no significant difference between the usage of superplasticizer A1 and A2. Based on this experimental work, the following recommendations can be proposed to further enhance the utility of the experiment. Different types of admixtures react differently in contact with cement, even if they are classified in the same category. Therefore, a study should be conducted to determine which admixture performs better under specific exposure conditions. In addition, only three different dosages of two different admixtures were used in this experiment. Therefore, it is difficult to determine the exact optimal dosage of the studied admixtures. For this reason, more concrete mixtures should be made with recycled rubber containing different dosages of admixtures to obtain the optimal dosage of the admixture.

ACKNOWLEDGEMENT

The research presented in this article is a part of the research project Development of Reinforced Concrete Elements and Systems with Waste Tire Powder – ReCoTiP (UIP-2017-

05-7113), supported by Croatian Science Foundation and its support is gratefully acknowledged.

REFERENCES

[1]   Emiroglu, M., Yildiz, S., Kele stemur, O. & Kele stemur, M.H., Bond performance of rubber particles in the self-compacting concrete. *Proceedings of the 4th International Symposium Bond in Concrete 2012: Bond, Anchorage, Detailing*, Brescia, Italy, 17–20 Jun., pp. 779–785, 2012.

[2]   Ismail, M.K. & Hassan, A.A.A., Use of metakaolin on enhancing the mechanical properties of self-consolidating concrete containing high percentages of crumb rubber. *J. Clean. Prod.*, **125**, pp. 282–295, 2016.

[3]   Gesoglu, M., Güneyisi, E., Khoshnaw, G., & Ipek, S., Investigating properties of pervious concretes containing waste tire rubbers. *Constr. Build. Mater.*, **63**, pp. 206–213, 2014 and *Journal of Agricultural Science and Engineering*, **1**(2), pp. 70–74, 2015.

[4]   Bušić, R., Benšić, M., Miličević, I. & Strukar, K., prediction models for the mechanical properties of self-compacting concrete with recycled rubber and silica fume. *Materials*, **13**(8), pp. 1–25, 2020. DOI: 10.3390/ma13081821.

[5]   Šandrk Nukić, I. & Miličević, I., Fostering eco-innovation: Waste tyre rubber and circular economy in Croatia. *Interdisciplinary Description of Complex Systems: INDECS*, **17**(2-B), pp. 326–344, 2019. DOI: 10.7906/indecs.17.2.9.

[6]   Bušić, R., Miličević, I., Kalman Šipoš, T. & Strukar, K., Recycled rubber as an aggregate replacement in self-compacting concrete: Literature overview. *Materials*, **11**(9), pp. 1–25, 2018. DOI: 10.3390/ma11091729.

[7]   Hadzima-Nyarko, M., Nyarko, E.K., Ademović, N., Miličević, I. & Kalman Šipoš, T., Modelling the influence of waste rubber on compressive strength of concrete by artificial neural networks. *Materials*, **12**(4), pp. 1–18, 2019. DOI: 10.3390/ma12040561.

[8]   Neville, A.M., *Properties of Concrete*, Pearson, Prentice Hall, pp. 255–262, 2005.

[9]   Alsadey, S., Effect of superplasticizer on fresh and hardened properties of concrete. *Journal of Agricultural Science and Engineering*, **1**(2), pp. 70–74, 2015.

[10]   Adams, R.F., Stodola, P.R. & Mitchell, D.R., Discussion of concrete retempering studies by Hawkins M.J. *ACI Journal Proceedings*, **59**, pp. 1249–1250, 1962.

[11]   Ravina, D., Retempering of prolonged-mixed concrete with admixtures in hot weather. *ACI Journal Proceedings*, **72**(6), pp. 291–295, 1975.

[12]   Previte, R.W., Concrete slump loss. *ACI Journal Proceedings*, 74(8), pp. 361–367, 1977.

[13]   Ravina, D. & Soroka, I., Slump loss and compressive strength of concrete made with WRR and HRWR admixtures and subjected to prolonged mixing. *Cement and Concrete Research*, **24**(8), pp. 1455–1462, 1994.

[14]   Baskoca, A., Ozkul, M.H. & Artirma, S., Effect of chemical admixtures on workability and strength properties of prolonged agitated concrete. *Cement and Concrete Research*, **28**(5), pp. 737–747, 1998.

[15]   Mohammed, T.U., Ahmed, T., Mallick, T.A., Shahriar, F. & Munim, A., Influence of chemical admixtures on fresh and hardened properties of ready mix concrete. *1st International Conference on Engineering Research Practice*.

[16]   Rahman, A., Mashiri, F., Rahman, M.M. & Karim, R. (eds), pp. 36–41, STAMCA: Dhaka, Bangladesh, 2017.

SECTION 2
STRUCTURAL OPTIMIZATION

OPTIMIZATION OF STEEL AND TIMBER HALL STRUCTURES

STOJAN KRAVANJA & TOMAŽ ŽULA
Faculty of Civil Engineering, Transportation Engineering and Architecture, University of Maribor, Slovenia

ABSTRACT

The paper deals with the optimization of single-storey hall structures consisting of the same main frames to which steel purlins, façade rails and façade columns are connected. The frames can be steel or timber portal frames. While the steel frames are made of steel I-sections, the timber frames are made of glulam with rectangular cross-sections. The hall structure is optimized using mixed-integer nonlinear programming (MINLP), a combined continuous-discrete optimization technique. MINLP optimization is performed in three steps. It starts with defining the hall superstructure, modelling the optimization model of the structure, and solving the defined optimization problem. The superstructure includes all discrete alternatives of topologies, standard dimensions and material qualities competing for a feasible and optimal result. The optimization model includes continuous and discrete binary variables. The continuous variables represent dimensions, cross-sections, material grades, loads, etc., while the binary variables are used to optimize the topology of the structure and to select standard dimensions/profiles and material grades. The objective function of the material cost of the structure is subject to a system of (in)equality constraints of structural analysis and dimensioning. The dimensioning constraints are defined according to the Eurocode regulations. In order to solve the defined optimization problem, the modified outer-approximation/equality-relaxation (OA/ER) algorithm was used. A numerical example of MINLP optimization of a steel and timber frame hall structure is presented at the end of the article.

Keywords: steel hall, timber hall, steel structures, timber structures, optimization, mixed-integer non-linear programming, MINLP.

1 INTRODUCTION

The optimization of steel and timber frames/hall structures represents a modern field within structural optimization. To achieve optimal frame design, researchers have developed a number of useful optimization methods suitable for both continuous and discrete optimization. O'Brien and Dixon [1] proposed a linear programming approach for optimal portal frame design. Guerlement et al. [2] presented a practical method in which they minimized the mass of a steel hall using Eurocode 3 [3]. Saka [4] and McKinstray et al. [5] calculated the optimal steel frame design using a genetic algorithm. Using mixed integer nonlinear programming, MINLP, Kravanja and Žula [6] optimized the production cost of the structure of a steel hall. Recently, Kravanja et al. [7] presented the parametric optimization of steel industrial halls. One of the latest research contributions in this field is the work of McKinstray et al. [8], in which the authors achieve the optimal shape of the main steel frame with minimal mass.

According to the number of available references, the area of optimization of hall structures with timber frames is less frequently discussed than the area with steel halls. In this area, Topping and Robinson [9] have performed optimization of timber frames with sequential linear programming and Kravanja and Žula performed optimization of hall structures with timber portal frames with mixed integer nonlinear programming, MINLP [10], [11].

The paper deals with the optimization of a single-storey hall structure with steel or timber main frames (see Fig. 1). The mentioned hall structures are built for industrial, sports and commercial purposes. We optimize the hall structures with mixed-integer nonlinear programming, MINLP. MINLP is a discrete/continuous method of mathematical

WIT Transactions on The Built Environment, Vol 209, © 2022 WIT Press
www.witpress.com, ISSN 1743-3509 (on-line)
doi:10.2495/HPSU220041

programming. MINLP contains both continuous and discrete variables. Continuous variables are defined for calculating continuous parameters, and discrete variables are used for discrete decision making. In MINLP, continuous and discrete optimization take place simultaneously. The results of such optimization are optimal continuous parameters (the optimal mass or production cost of the structure) and discrete parameters (the optimal topology of the structure, the strength classes of the different materials, and the discrete standard/rounded dimensions).

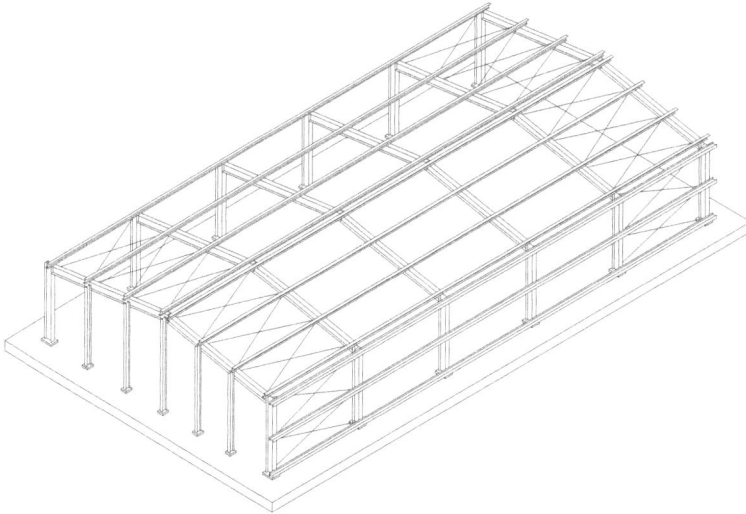

Figure 1: Structure of a steel hall.

2 MINLP PROBLEM FORMULATION

The optimization problem of hall structures is non-linear, non-convex, continuous, and discrete. Therefore, MINLP is chosen for the optimization. It is assumed that a general non-convex, non-linear, discrete, and continuous optimization problem can be formulated as a MINLP problem in the form of:

$$\min z = c^T y + f(\mathbf{x})$$

subjected to:

$$h(x) = 0 \qquad\qquad \textbf{(MINLP)}$$

$$g(x) \leq 0$$

$$By + Cx \leq b$$

$$x \in X = \{x \in R^n : x_{LO} \leq x \leq x_{UP}\}$$

$$y \in Y = \{0,1\}^m$$

The above mathematical MINLP formulation contains the objective function z, the nonlinear functions $f(x)$, $h(x)$ and $g(x)$, the mixed linear equality/inequality constraints $By + Cx \leq b$, the vector of continuous variables x and the vector of discrete binary variables y.

In structural optimization, the continuous variables x define the dimensions, strains, stresses, costs etc., while the binary variables y represent the potential existence of structural elements within the defined superstructure and the choice of discrete/standard materials and sizes. Non-linear equality and inequality constraints and the bounds of the continuous variables represent the rigorous system of design, loading, resistance, and deflection constraints known from structural analysis.

The general MINLP model formulation has been adapted for the optimization of mechanical superstructures (MINLP-SMS). The resulted formulation is more specific, particularly in variables and constraints. It was used also for the modelling of steel and timber framed structures, see also Žula and Kravanja [12]. This formulation is given in the following form:

$$\min z = c^{\mathrm{T}}y + f(\mathbf{x})$$

subjected to:

$$h(\mathbf{x}) = \mathbf{0}$$

$$g(\mathbf{x}) \leq \mathbf{0}$$

$$Ey \leq e \qquad\qquad\qquad\qquad \textbf{(MINLP-SMS)}$$

$$Dy^e + R(\mathbf{x}) \leq r$$

$$Ky^e + L(d^{cn}) \leq k$$

$$Py + S(d^{st}) \leq s$$

$$Ay + B(d^{mat}) \leq a$$

$$x \in X = \{x \in R^n : x\mathrm{LO} \leq x \leq x\mathrm{UP}\}$$

$$y \in Y = \{0,1\}^m$$

In the model formulation, included are continuous variables $x = \{d, p\}$ and discrete binary variables $y = \{y^e, y^{st}, y^{mat}\}$. The continuous variables are partitioned into design variables $d = \{d^{cn}, d^{st}, d^{mat}\}$ and into performance (non-design) variables p, where sub-vectors d^{cn}, d^{st} and d^{mat} stand for the continuous dimension, standard dimensions and standard material strengths, respectively. Sub-vectors of binary variables y^e, y^{st} and y^{mat} denote the potential existence of structural elements inside the superstructure (the topology determination), the potential selection of standard dimension alternatives and standard material strengths, respectively.

The economical objective function z involves fixed cost charges in the linear term $c^{\mathrm{T}}y$ and dimension dependent costs in the term $f(\mathbf{x})$.

The parameter nonlinear and linear constraints $h(\mathbf{x}) = \mathbf{0}$ and $g(\mathbf{x}) \leq \mathbf{0}$ represent the rigorous system of the design, loading, resistance, stress, deflection, etc., constraints known from the structural analysis.

The integer linear constraints $Ey \leq e$ are proposed to describe the relations between binary variables.

The mixed linear constraints $Dy^e + R(\mathbf{x}) \leq r$ restore interconnection relations between currently selected or existing structural elements (corresponding $y^e = 1$) and cancel relations for currently disappearing or non-existent elements (corresponding $y^e = 0$).

The mixed linear constraints $Ky^e + L(d^{cn}) \leq k$ are proposed to define the continuous design variables for each existing structural element. The space is defined only when the corresponding structure element exists ($y^e = 1$), otherwise it is empty.

The mixed linear constraints $Py + S(d^{st}) \leq s$ define the standard design variables d^{st}. Each standard dimension d^{st} is determined as a scalar product between its vector of standard/discrete dimension constants q and its vector of binary variables y^{st}, see eqn (1). Only one discrete value can be selected for each standard dimension, see eqn (2).

The mixed linear constraints $Ay + B(d^{mat}) \leq a$ determine the standard material strengths of steel, timber and concrete. These conditions are defined in a similar way as the standard design variables (sections). For more on the MINLP model formulation and how to handle the equations and variables, see Kravanja et al. [13]–[15].

$$d^{st} = \sum_{i \in I} q_i y_i^{st} \tag{1}$$

$$\sum_{i \in I} y_i^{st} = 1 \tag{2}$$

The hall structure under study consists of the same main frames on which the roof purlins, façade rails, and front and rear façade columns are connected. Purlins, rails and façade columns are made of hot-rolled IPE or HEA steel sections. The main frames may be steel or timber portal frames. The steel frames are made of HEA profiles and the timber frames are made of glued laminated timber with rectangular cross-section. The columns of the portal frames are supported by concrete pad foundations. The MINLP superstructure of the hall is defined, which presents a variety of different topological/structural alternatives with:

- sets and binary variables for topology alternatives of frames, purlins, rails and façade columns;
- sets and binary variables for standard-dimension alternatives of steel profiles, timber cross-sections and concrete pad foundation;
- sets and binary variables for standard strength classes of different materials.

3 OPTIMIZATION MODELS

The MINLP optimization models HALLOPT (HALL OPTimization) were developed for the optimization of hall structures, an extra model for the hall with steel frame and an extra model for the hall with timber frames. These models were developed based on the presented MINLP model formulation. The Optimization models include input data (constants), variables, and a cost objective function subject to load, stress, resistance, and deflection (in)equality constraints for dimensioning, and integer logical constraints for topology, standard dimensions, and strength class calculations.

The optimization of the considered structure is performed by the combined action of the self-weight of the frame elements, the vertical uniformly distributed variable load (snow) and the horizontal concentrated variable load located at the top of the columns (wind). The internal forces are calculated according to the first-order elastic theory. Steel frames are defined as non-sway frames ($\alpha_{cr} \geq 10$) and timber frames are treated as sway frames. Longitudinal stability is provided by a bracing system. Eurocode 3 [3] has been used for the dimensioning of structural steel elements, where all conditions for ultimate limit state (ULS) and serviceability limit state (SLS) are fulfilled. For ULS, the elements are checked for the following:

- axial force;
- shear force;
- bending moment;

- resistance to compression buckling;
- lateral-torsional buckling resistances;
- interaction between compression buckling resistance and lateral-torsional buckling resistance, see Kravanja et al. [7].

Eurocode 5 [16] has been used for the dimensioning of a glulam frame. These equations include the cross-sectional resistances of the columns and beams for:

- axial compression force;
- bending moment;
- shear force;
- compression buckling resistance;
- lateral torsional stability;
- interaction between compression buckling resistance and lateral-torsional buckling resistance, see Kravanja et al. [7].

Stresses in the apex zone have been also checked, see Kravanja and Žula [11]. For SLS, the vertical deflections of steel and timber elements and the horizontal displacements of the main frames are checked.

Input data (constants) in the models include span, height and length of the hall structure, alternatives (strengths) of standard materials used (steel, timber, concrete), standard/discrete section alternatives, vertical load (snow), horizontal load (wind), weights of the roof and cladding, prices of steel, timber and concrete, safety factors, etc. The variables included in the models are continuous and discrete binary variables. The continuous variables represent the material costs of fabricating the structure, the number of main frames, purlins, and rails, the intermediate spacing between main frames, purlins, and rails, the section dimensions, areas, section moduli, torsional constants, etc. Note that the above mentioned (in)equality constraints for the ultimate and serviceability limit states (the axial and shear forces, bending moments, deflections, etc.) depend on the input data and the variables calculated.

3.1 Objective function

The material cost of the structure MAT_{COST} is defined as the sum of the material costs for the fabrication of the main frames, roof purlins, façade rails, façade columns, and foundations, see eqn (3):

$$MAT_{COST} = \{2 \cdot [(A_C \cdot H_C \cdot \rho_{FRAME}) \cdot NO_{FRAME} + (A_B \cdot L_B \cdot \rho_{FRAME}) \cdot NO_{FRAME}]\} \cdot ECM_{FRAME} \quad (3)$$

$$+ \{(A_P \cdot L_{TOT} \cdot \rho_{STEEL}) NO_{PURL} + (A_R \cdot L_{TOT} \cdot \rho_{STEEL}) \cdot NO_{RAIL}$$

$$+ 2 \cdot (A_{FC} \cdot H_{FC} \cdot \rho_{STEEL}) \cdot (NO_{PURL} {}^-1)\} \cdot ECM_{STEEL}$$

$$+ 2 \cdot (H_F \cdot B_F \cdot B_F) \cdot NO_{FRAME} \cdot ECM_{CONCR}$$

where A_C in eqn (3) is the cross-section of the frame column, H_C is the height of the column, ρ_{FRAME} is the volume density of the frame material (steel or glulam), NO_{FRAME} is the number of main frames, A_B is the cross-section of the frame beam, L_B is half the length of the inclined beam (approx. span/2), and ECM_{FRAME} is the unit price of steel or glulam frame material (€/kg). In addition, A_P is the cross-section of the roof purlin, L_{TOT} is the length of the hall, ρ_{STEEL} is the volume mass of the steel, NO_{PURL} represents the number of purlins, A_R is the cross-section of the façade rail, NO_{RAIL} represents the number of façade rails, A_{FC} is the cross-section of the front and rear façade columns, H_{FC} stands for the height of the façade column,

WIT Transactions on The Built Environment, Vol 209, © 2022 WIT Press
www.witpress.com, ISSN 1743-3509 (on-line)

ECM_{STEEL} is the unit price of steel (€/kg), H_F is the height of the square pad foundations, B_F is the width of the foundations and ECM_{CONCR} is the unit price of concrete (€/m³). The mentioned numbers of structural elements and cross-sections are defined as variables that are calculated when the objective function MAT_{COST} converges.

4 NUMERICAL EXAMPLE

In the paper, the numerical example of MINLP optimization of a hall structure with a span of 16 m, a length of 80 m and a height of 4.5 m is presented. The structure of the hall is loaded with its own weight, the weight of the roofing 0.20 kN/m², the weight of the façade cladding 0.15 kN/m², snow 0.80 kN/m² and with the horizontal wind 0.50 kN/m². The unit price of the steel is 1.5 €/kg, the price of the glulam is 1000 €/m³ and that of the concrete is 150 €/m³.

The MINLP optimization models of a single-storey steel and timber hall structures are modelled in the GAMS (General Algebraic Modelling System) environment [17]. The defined MINLP optimization problem is solved using the Modified Outer-Approximation/Equality-Relaxation algorithm of Kravanja and Grossmann [18], see also Kravanja et al. [13]. The algorithm works as an alternative sequence of non-linear programming (NLP) and mixed-integer linear programming (MILP) subproblems. The MINLP computer program MIPSYN [19] is applied for the optimization. GAMS/CONOPT (generalized reduced-gradient method) [20] is used for NLP continuous calculations and GAMS/CPLEX (branch and bound) [21] for MILP discrete optimizations.

4.1 Hall structure with steel frames

In the optimization model HALLOPT, we defined the superstructure of the steel hall, which represents variety of different alternatives of structural elements, cross-sections and strength classes of materials: 70 portal frames, 2 × 25 purlins, 2 × 5 façade rails, 18 hot rolled IPE profiles (from IPE 80 to IPE 600, extra for purlins and rails), 24 hot-rolled HEA sections (from HEA 100 to HEA 1000, extra for frame columns, frame beams, and the front and rear wall façade columns), three structural steel grades (S 235, S 275 and S 355) and 121 different discrete alternative widths for the 1.20 m deep square foundations (from 50 to 350 cm, in 2.5 cm increments). The defined superstructure of the hall contains $1.4226 \cdot 10^{13}$ different alternative structures – one of which is the optimal one.

The MINLP optimization was carried out using the computer program MIPSYN, which required a working time of about 20 minutes for the calculation. The optimal result of €100,245 was found in the 19th major MINLP iteration, see the convergence in Table 1.

Note that the OA/ER algorithm consists of an alternative sequence of non-linear programming (NLP) optimization subproblems and mixed-integer linear programming (MILP) main problems. The discrete optimization MILP involves a global linear approximation to the structure and predicts a new set of binary variables, i.e., a new topology and standard dimensions/materials. The NLP subproblem corresponds to the continuous optimization of the structure for the computed topology and standard dimensions/materials given at the corresponding MILP. From iteration to the iteration (MILP and NLP), the NLPs get better results, and the MILPs get worse results. The search is terminated when the NLP matches the MILP result. In the case of the non-convex problem, the search is stopped when the NLP can no longer be improved.

A three-phase MINLP strategy was used for the optimization. The first phase corresponds to the initialization, in which the continuous NLP optimization is performed. The first NLP solution of €84,308 is used as a good starting point for further discrete optimization. The

Table 1: Convergence to the optimal result for the steel hall.

MINLP iteration	MINLP subphase	Result €
Phase 1: Continuous optimization		
1	Initialization 1.NLP	*84,308*
Phase 2: Discrete topology and material optimization		
2	1.MILP 2.NLP	85,142 *85,475*
3	2.MILP 3.NLP	85,484 86,546
Phase 3: Discrete topology, material and standard dimension optimization		
4	3.MILP 4.NLP	2,068,470 *98,664* loc. infeas.
. . .	. . .	. . .
19	18.MILP 19.NLP	2,070,029 **100,245**
20	19.MILP 20.NLP	2,070,531 100,973 loc. infeas.
. . .	. . .	. . .
25	24.MILP 25.NLP	2,070,549 100,788

second phase corresponds to the simultaneous topology and material discrete optimization. The dimensions/sections are still continuous (not discrete) in this phase. The optimal solution of the second phase is found in the 2nd MINLP iteration and yields €85,475. Since this problem is highly non-linear and non-convex optimization problem, the search is terminated if the NLP does not yield an improvement (the next 3rd iteration shows a worse solution of €86,546). After the optimal solution of the second phase is calculated, the optimization continues with the overall topology, material and standard dimension optimization of the structure at the third phase. In this phase, the best NLP solution of €100,245 is found in the 19th iteration, since the first real subsequent NLP solution shows a worse result of €100,788 at the 25th iteration (note that all other subsequent solutions before the 25th iteration are locally infeasible).

The optimal result represents the lowest possible material cost of the hall structure, namely €100,245, the optimal number of 16 portal frames, 12 purlins and eight rails. Obtained are the optimal profiles HEA 280 for the frame columns, HEA 320 for the frame beams, IPE 120 for the purlins, IPE 140 for the rails, HEA 120 for the façade columns and the square concrete pad foundations $1.20 \times 1.675 \times 1.675$ m^3. Steel class S 355 is calculated (see Fig. 2).

4.2 Hall structure with timber frames

The second calculation deals with the optimization of the same hall structure as above, but the main frames are made of a glulam with a rectangular cross-section. The paper deals with

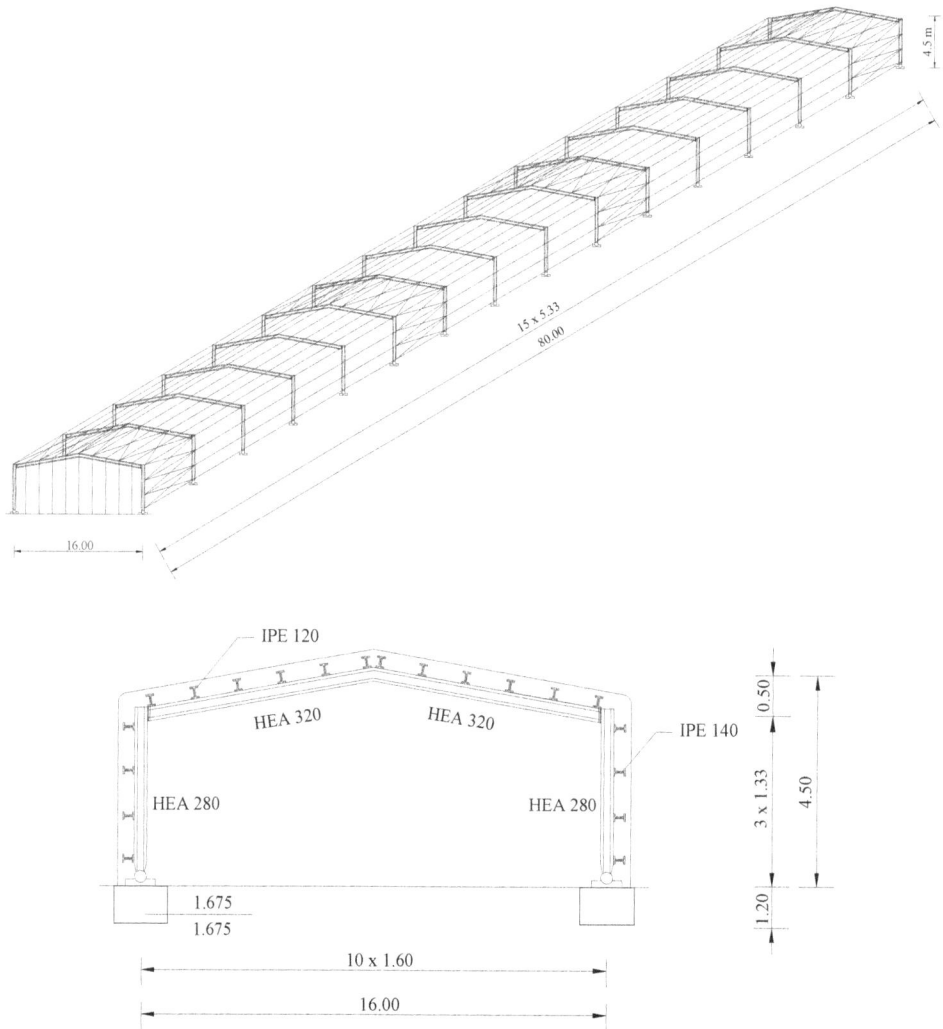

Figure 2: Optimal structure of the hall with main steel frames.

a special case, where the columns and beams of the timber frames have the same cross-section. The superstructure of the hall contains 70 portal frames, 2 × 25 purlins, 2 × 5 rails, 17 discrete alternatives for the width of the timber rectangular cross-section of the frame (from 10 to 50 cm, in 2.5 cm steps), 51 alternatives for the height of the timber frame cross-section (from 25 to 150 cm, in 2.5 cm steps), 18 hot-rolled IPE profiles (from IPE 80 to IPE 600 for purlins and rails separately), 24 hot-rolled HEA sections (from HEA 100 to HEA 1000 for façade columns), three structural steel classes (S 235, S 275 and S 355) and 121 different discrete alternative widths for the 1.20 m deep square foundations (from 50 to 350 cm, 2.5 cm increments). The material of the glulam is GL28h. The superstructure of the hall contains $6.4240 \cdot 10^{13}$ different structural alternatives – one of them is optimal.

The minimal calculated material cost of the hall structure with main frame made of timber resulted in €138,207. The computer program MIPSYN needed a working time of about 15 minutes for the calculation. The optimal result was found in the 9th major MINLP iteration, see the convergence in Table 2. The optimal number of 12 portal frames, 12 purlins and eight rails was calculated. The optimal cross-sectional dimensions of the timber frames are 200/1,200 mm². The optimal profiles IPE 160 for the purlins, IPE 200 for the rails and HEA 120 for the façade columns were determined (see Fig. 3). The dimensions of the square concrete foundation are $1.20 \times 1.925 \times 1.925$ m³. The steel grade S 355 was calculated.

Table 2: Convergence to the optimal result for the timber hall.

MINLP iteration	MINLP subphase	Result €
Phase 1: Continuous optimization		
1	Initialization	
	1.NLP	*123,279*
Phase 2: Discrete topology and material optimization		
2	1.MILP	96,712
	2.NLP	129,021
3	2.MILP	101,529
	3.NLP	125,370
4	3.MILP	104,320
	4.NLP	*124,990*
5	4.MILP	105,772
	5.NLP	127,602
Phase 3: Discrete topology, material and standard dimension optimization		
6	5.MILP	2,954,022
	6.NLP	*137,290* loc. infeas.
7	6.MILP	2,954,455
	7.NLP	*137,745* loc. infeas.
8	7.MILP	2,954,615
	8.NLP	*137,884* loc. infeas.
9	8.MILP	2,954,889
	9.NLP	**138,207**
10	9.MILP	2,955,049
	10.NLP	138,377

5 CONCLUSIONS

The paper deals with the optimization of the structure of a single-storey hall consisting of the same main frames to which steel purlins, façade rails and façade columns are connected. The frames may be made of steel profiles or glulam. The optimization of the hall structure is performed using mixed-integer non-linear programming, MINLP. The objective function of the material cost of the structure is subject to a system of (in)equality constraints of statics and dimensioning. The modified outer-approximation/equality-relaxation algorithm (OA/ER) is applied to solve the optimization problem. The computer program MYPSIN is used. In addition to the determined minimal material cost of the structure, the optimal topology of the hall structure, the strength classes of the materials used, the standard steel profiles, and the discrete/rounded cross-sections of the glulam frames and of the concrete foundations are calculated.

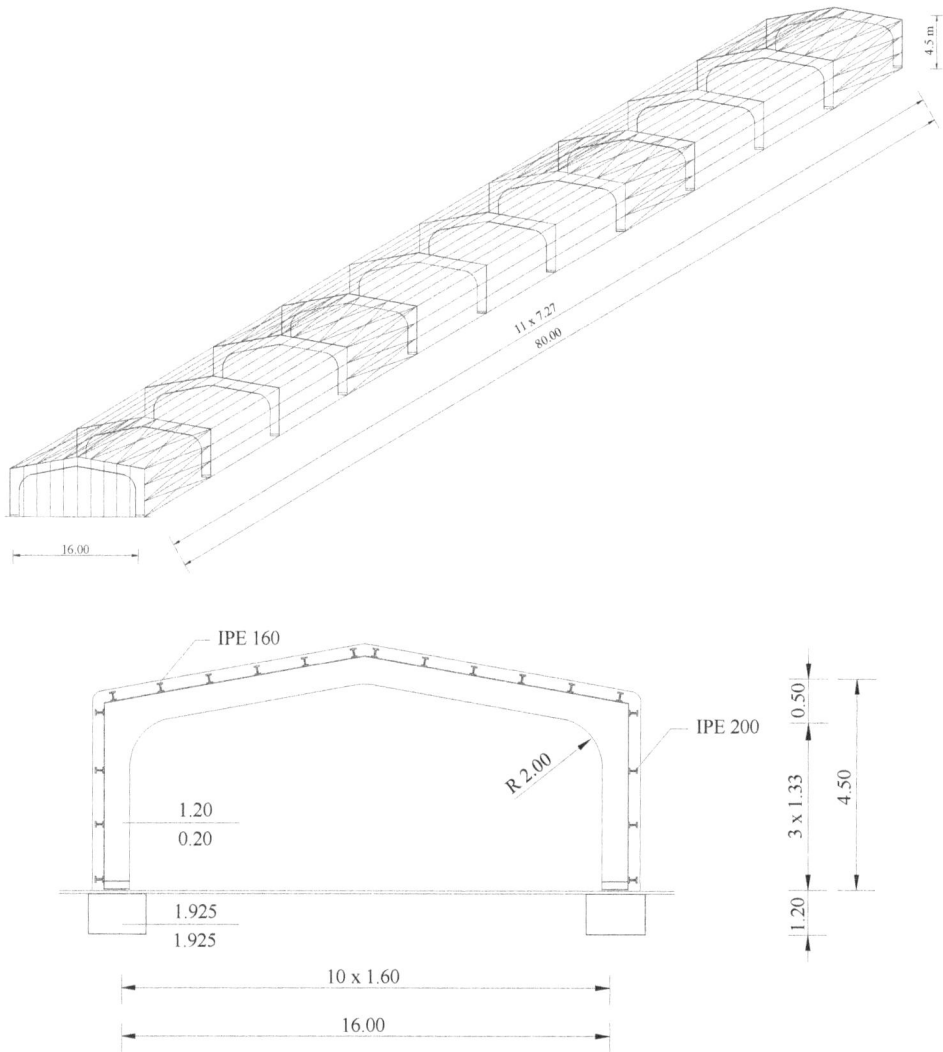

Figure 3: Optimal structure of the hall with main timber frames.

A numerical example of MINLP optimization of a hall structure is presented at the end of the article. For the given span, load, unit prices of steel, timber and concrete, we calculated the minimal material cost of the hall, separately for the steel-framed hall type and separately for the timber-framed hall version. Such cost optimization is very useful for quick comparison and selection of the optimal construction variant. Note that the steel frame hall version turned out to be more advantageous than the timber frame version, especially because of the high price of the glulam used.

ACKNOWLEDGEMENT

The authors are grateful for the support of funds from the Slovenian Research Agency (program P2-0129).

REFERENCES
[1]   O'Brien, E.J. & Dixon, A.S., Optimal plastic design of pitched roof frames for multiple loading. *Computers and Structures*, **64**, pp. 737–740, 1997.
[2]   Guerlement, G., Targowski, R., Gutkowski, W., Zawidzka J. & Zawidzki, J., Discrete minimum weight design of steel structures using EC3 code. *Structural and Multidisciplinary Optimization*, **22**, pp. 322–327, 2001.
[3]   Eurocode 3, *Design of Steel Structures*, European Committee for Standardization: Brussels, 2005.
[4]   Saka, M.P., Optimum design of pitched roof steel frames with haunched rafters by genetic algorithm. *Computers and Structures*, **81**, pp. 1967–1978, 2003.
[5]   McKinstray, R., Lim James, B.P., Tanyimboh Tiku, T., Phanc Duoc T. & Sha W., Optimal design of long-span steel portal frames using fabricated beams. *Journal of Constructional Steel Research*, **104**, pp. 104–114, 2015.
[6]   Kravanja, S. & Žula, T., Cost optimization of industrial steel building structures. *Advances in Engineering Software*, **41**(3), pp. 442–450, 2010.
[7]   Kravanja, S., Turkalj, G., Šilih, S. & Žula, T., Optimal design of single-story steel building structures based on parametric MINLP optimization. *Journal of Constructional Steel Research*, **81**, pp. 86–103, 2013.
[8]   McKinstray, R., Lim James, B.P., Tanyimboh Tiku, T., Phanc Duoc, T. & Sha W., Comparison of optimal designs of steel portal frames including topological asymmetry considering rolled, fabricated and tapered sections. *Engineering Structures*, **111**, pp. 505–524, 2016.
[9]   Topping, B.H.V. & Robinson, D.J., Optimization of timber framed structures. *Computers and Structures*, **18**(6), pp. 1167–1177, 1984.
[10]   Kravanja, S. & Žula, T., Optimization of a timber hall structure. *High Performance and Optimum Design of Structures and Materials IV, WIT Transactions on the Built Environment,* vol. 196, WIT Press: Southampton and Boston, pp. 183–192, 2020.
[11]   Kravanja, S. & Žula, T., Optimization of a single-storey timber building structure. *International Journal of Computational Methods and Experimental Measurements*, **9**(2), pp. 126–140, 2021.
[12]   Žula, T. & Kravanja, S., The two-phase MINLP optimization of a single-storey industrial steel building. *High Performance Structures and Materials IV, WIT Transactions on the Built Environment,* vol. 97, WIT Press: Southampton and Boston, pp. 439–448, 2008.
[13]   Kravanja, S., Kravanja, Z. & Bedenik, B.S., The MINLP optimization approach to structural synthesis. Part I: A general view on simultaneous topology and parameter optimization. *International Journal of Numerical Methods in Engineering*, **43**, pp. 263–292, 1998.
[14]   Kravanja, S., Kravanja, Z. & Bedenik, B.S., The MINLP optimization approach to structural synthesis. Part II: Simultaneous topology, parameter and standard dimension optimization by the use of the Linked two-phase MINLP strategy. *International Journal of Numerical Methods in Engineering*, **43**, pp. 293–328, 1998.
[15]   Kravanja, S., Kravanja, Z. & Bedenik, B.S., The MINLP optimization approach to structural synthesis. Part III: Synthesis of roller and sliding hydraulic steel gate structures. *International Journal of Numerical Methods in Engineering*, **43**, pp. 329–364, 1998.
[16]   Eurocode 5, *Design of Timber Structures*, European Committee for Standardization: Brussels, 2008.

[17] Brooke, A., Kendrick, D. & Meeraus, A., *GAMS: A User's Guide*, Scientific Press: Redwood City, CA, 1988.
[18] Kravanja, Z. & Grossmann, I.E., New developments and capabilities in PROSYN: An automated topology and parameter process synthesizer. *Computers and Chemical Engineering*, **18**, pp. 1097–1114, 1994.
[19] Kravanja, Z., Challenges in sustainable integrated process synthesis and the capabilities of an MINLP process synthesizer MIPSYN. *Computers and Chemical Engineering*, **34**(11), pp. 1831–1848, 2010.
[20] Drudd, A.S., CONOPT: A large-scale GRG code. *ORSA Journal on Computing*, **6**, pp. 207–216, 1994.
[21] CPLEX User Notes, ILOG Inc.

OPTIMUM DESIGN OF CABLE-STAYED BRIDGES CONSIDERING CABLE FAILURE

NOEL SOTO, CLARA CID, AITOR BALDOMIR & SANTIAGO HERNÁNDEZ
Structural Mechanics Group, School of Civil Engineering, University of A Coruña, Spain

ABSTRACT
A methodology to obtain the minimum weight of cables in cable-stayed bridges when a cable fails has been developed. To this end, a multi-model strategy is proposed that takes into account design constraints in both the intact and damaged models. The dynamic effect of the cable breakage is considered by the application of impact loads at the tower and deck anchorages. The methodology is applied to the Queensferry Crossing Bridge, a multi-span cable-stayed bridge with cross stay cables in the central section of each main span. The number of cables, anchorage position on the deck, cable areas and prestressing forces are considered as design variables into the optimization process simultaneously. The fail-safe optimum design results in a different cable layout than the optimized design of the intact structure, with minimum volume increase.
Keywords: cable-stayed bridge, optimum design, fail-safe, cable rupture, cable breakage, cable system, cable arrangement, cable layout.

1 INTRODUCTION

Optimization techniques applied to cable-stayed bridges have gained prominence in the research community. While several papers focus on optimizing the shape or thicknesses of the deck and cable areas [1]–[8], other researchers have concentrated their efforts on minimizing the weight and arrangement of the cable system. The reason is that a reduction in the steel volume of the cable system can lead to considerable savings, since the cable system represents approximately 10% of the total cost of the bridge, as presented in Sun et al. [9]. In this sense, the determination of the optimum cable forces distribution has been thoroughly studied [10]–[13]. Among these works, Baldomir et al. [12] obtained the cable areas for a long span bridge by minimizing the cables volume through a gradient-based optimization algorithm. Then, Baldomir et al. [14] considered a multi-model optimization technique to minimize the cable weight with crossing cables and fixed anchor positions. Cid et al. [15] proposed a methodology to define the optimum cable system in multi-span cable-stayed bridges, allowing crossed cables in the main spans, different number of cables at each side of the towers and different cable areas. Martins et al. [16] presented a comprehensive summary of the state-of-the-art through an extensive literature survey, with 90 articles studied for a detailed review.

The previous approaches were applied to the intact configuration of the bridge. Therefore, a weakness of those optimum designs is that they do not contemplate a cable break scenario. As several dramatic events have occurred throughout history associated with this kind of accidents, it seems appropriate to propose a strategy to optimize the cable system that takes into account a cable failure. Cross sectional areas, cable anchor positions and post-tensioning cable forces will be the design variables of the problem. A MATLAB code [17] has been programmed and combined the structural analysis software ABAQUS [18] in order to solve the proposed fail-safe optimization problem.

WIT Transactions on The Built Environment, Vol 209, © 2022 WIT Press
www.witpress.com, ISSN 1743-3509 (on-line)
doi:10.2495/HPSU220051

2 OPTIMIZATION STRATEGY

2.1 Structural analysis considering a cable loss

The existing codes and regulations in civil engineering field establish that bridges must resist a single-cable breakage. The structural response derived from this accidental event can be contemplated by non-linear dynamic analyses or by a quasi-static approach. The quasi-static approach assumes that two impact forces must be applied in the opposite direction of the broken cable. These static forces correspond to the cable tensile strength multiplied by an amplification factor, denoted as DAF, which can be understood as the ratio between the dynamic response and the static response [19]. Eurocodes and the PTI recommendations establish a value the DAF between 1.5 and 2.0. Fig. 1 shows the impact load due to the loss of a cable.

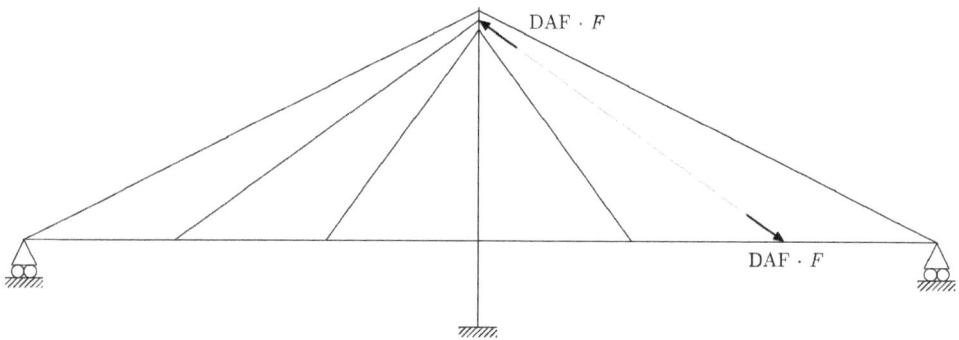

Figure 1: Impact load due to the loss of a cable.

The load combination used for the cable loss can be found in the PTI recommendations [20] and it is presented in eqn (1), being DC and DW the dead load of structural and non-structural components, respectively, LL the live load, and $CLDF$ the cable loss dynamic forces.

$$1.1 \cdot DC + 1.35 \cdot DW + 0.75 \cdot LL + 1.1 \cdot CLDF. \qquad (1)$$

Thus, the proposed methodology will integrate the cable loss effect in the fail-safe optimization approach presented below.

2.2 Formulation of the optimization problem

The objective is to minimize the total steel volume of material of the cable system when a cable breaks. The design variables are the anchorage position on the deck (x_k^P), the cross-sectional area of the cables (x_k^A), the number of cables and their prestressing forces (x_k^F).

Since there could be as many damaged bridge configurations as there are cables, a multi-model optimization should be considered. This idea was proposed by Baldomir et al. [21] to achieve safe designs with minimum weight while fulfilling all limit-state requirements for the intact model and a set of partial collapses. It is therefore an optimization problem with a high computational cost since the design constraints must be evaluated by carrying out structural analyses in the intact and all possible damaged configurations.

A general formulation of the optimization problem is presented as follows:

$$\min V = 2 \cdot \sum_{k=1}^{N}\left(x_k^A \cdot L_k(x_k^P) \right) \tag{2a}$$

s.t.

$$g_j^{Mi}(x_k^P, x_k^A, x_k^F) \leqslant 0 \quad k = 1, \dots, N \quad j = 1, \dots, m^{M_i} \quad i = 0, \dots, D \tag{2b}$$

where V is the total volume of steel in the cable system; L_k is the total length of the cable k; and N is the number of cables. The whole set of design constraints is represented by the expressions $g_j^{Mi,l} \leq 0$, where M_i refers to the structural configuration i. If $i = 0$ the constraint refers to the intact model, while if it is non-zero, it refers to a damaged model.

As can be seen, the number of cables has not been explicitly considered as a design variable. In fact, it should be considered as a binary variable, whose value would be 0 if the cable did not exist and 1 if it did. Such an approach would entail the use of optimization algorithms with discrete variables that have proven to be inefficient when the number of variables is high. In this research, the existence or non-existence of the cable is considered as a function of its area, i.e., by means of a continuous variable. A lower limit of the cable area is defined with a very low value and if the variable tends to this value, the cable will be considered not to exist. By doing so, all the design variables of the optimization problem are continuous and a gradient-based optimization algorithm could be used to solve the problem presented in eqn (2).

The optimization code was implemented in MATLAB. After defining the mechanical properties, geometry, mesh of FEM, design variables and optimization parameters, the Python Script generates $D+1$ finite element models of the bridge. The sequential quadratic programming (SQP) algorithm implemented in the MATLAB function *fmincon* was used as optimizer. The Python Script is externally run through Abaqus at each iteration of the optimization process to obtain the structural responses and evaluate the design constraints. The intact model has to be analyzed first in order to obtain the internal forces of the cables. Then, the impact forces are applied to the damaged models, which are launched in parallel to evaluate their design constraints. The process is repeated until the objective function converges and the design constraints are satisfied.

3 APPLICATION EXAMPLE

3.1 Bridge description

The previous optimization strategy will be applied to the Queensferry Crossing Bridge, also known as Forth Replacement Crossing. A view of the bridge is shown in Fig. 2. The cable-stayed bridge has three towers around 200 m high, with a deck 1,950 m long, divided into two main spans of 650 m and two lateral spans of 325 m, being the latter composed of a back span of 221 m and one approach viaduct of 104 m. A scheme of the bridge is presented in Fig. 3.

The number of cables in the bridge is $N = 144$. As the FEM used is 2D, the cables of the model represent the combined capacity of the two cable planes of the real bridge. The mechanical properties of the deck and towers are summarized in Cid et al. [15]. The permanent loads applied to the FE model are the structural deck weight ($DC = 146$ kN/m), the weight of the non-structural elements of the deck ($DW = 54$ kN/m) and the cable prestressing forces (PS). It also was considered a live load on spans 1 and 3 ($LL1 = 102.5$ kN/m) and their symmetrical case on the spans 2 and 4 ($LL2$). Finally, cable loss dynamic forces ($CLDF$) are applied to damaged models. Load combinations considered in the optimization problem are shown in eqn (3).

Figure 2: Queensferry Crossing Bridge.

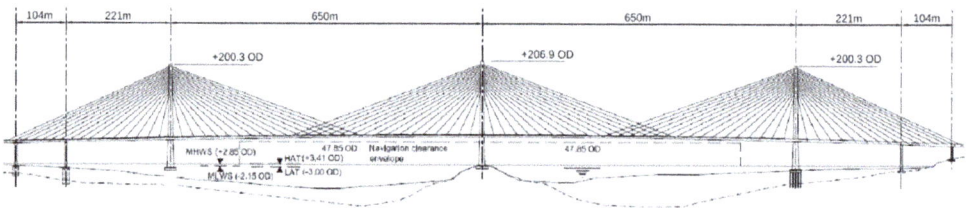

Figure 3: Scheme of the Queensferry Crossing Bridge [22].

Intact model:

Load Case 0 ($l = 0$) SLS: $1.00 \cdot DC + 1.00 \cdot DW + 1.00 \cdot PS$ (3a)

ULS: $1.25 \cdot DC + 1.50 \cdot DW + 1.25 \cdot PS$ (3b)

Load Case 1 ($l = 1$) SLS: $1.00 \cdot DC + 1.00 \cdot DW + 1.00 \cdot LL1 + 1.00 \cdot PS$ (3c)

ULS: $1.25 \cdot DC + 1.50 \cdot DW + 1.75 \cdot LL1 + 1.25 \cdot PS$ (3d)

Damaged models:

Load Case 1 ($l = 1$) EELS: $1.1 \cdot DC + 1.35 \cdot DW + 0.75 \cdot LL1 + 1.1 \cdot CLDF$ (3e)

Load Case 2 ($l = 2$) EELS: $1.1 \cdot DC + 1.35 \cdot DW + 0.75 \cdot LL2 + 1.1 \cdot CLDF$ (3f)

The volume of the cable-system in the real bridge corresponds to 759 m³. In a previous work [15], the optimization of the intact model was performed, resulting in a volume of 634.15 m³, which corresponds to a reduction of 16.45%.

3.2 Fail-safe optimization of the Queensferry Crossing Bridge

The formulation of the fail-safe optimization problem is presented in eqn (4):

$$\min V = 2 \cdot \sum_{k=1}^{N}\left(x_k^A \cdot L_k(x_k^P)\right) \tag{4a}$$

s.t

$$|w_j^{M0,l}| \leq w_{\max}^{M0,l} \qquad\qquad j = 1, \dots, N_D \qquad\qquad l = 0,1 \tag{4b}$$

$$|u_{\text{tower},p}^{M0,l}| \leq u_{\max}^{M0,l} \qquad\qquad p = 1,2,3 \qquad\qquad l = 0,1 \tag{4c}$$

$$0 < \sigma_{\text{cable},k}^{Mi,l} \leq \sigma_{\text{cable,max}} \qquad k = 1, \dots, N \ / \ k \neq i \quad l = 1,2 \quad i = 0, \dots, D \tag{4d}$$

$$\sigma_{C,\text{deck}} \leq \sigma_{\text{top,deck},j}^{Mi,l} \leq \sigma_{T,\text{deck}} \qquad j = 1, \dots, E_D \qquad l = 1,2 \quad i = 0, \dots, D \tag{4e}$$

$$\sigma_{C,\text{deck}} \leq \sigma_{\text{bottom,deck},j}^{Mi,l} \leq \sigma_{T,\text{deck}} \quad j = 1, \dots, E_D \qquad l = 1,2 \quad i = 0, \dots, D \tag{4f}$$

$$|x_{k+1} - x_k| \geq d_{\min} \qquad\qquad k = 1, \dots, N_P - 1 \tag{4g}$$

where:

$w_j^{M0,l}$ (m)		Deflection of the node j in the deck
$w_{\max}^{M0,l}$ (m)	$w_{\max}^{M0,0} = L/7500$ $w_{\max}^{M0,1} = L/500$	Maximum allowable deflection in the deck (side span: $L = 325$ m, Main span: $L = 650$ m)
N_D	$= 244$	Total number of deck nodes in which the displacements are checked
$\sigma_{\text{cable},k}^{Mi,l}$ (MPa)		Tensile stress in the cable k
$\sigma_{\text{cable max}}$ (MPa)	$= 837$	Maximum allowable tensile stress in cables
$\sigma_{\text{bottom,deck},j}^{Mi,l}$		Normal stress in the bottom fiber of the deck in jth element
$\sigma_{\text{top,deck},j}^{Mi,l}$		Normal stress in the top fiber of the deck in jth element
$\sigma_{C,\text{ deck}}$ (MPa)	$= -200$	Minimum allowable compression stress in deck
$\sigma_{T,\text{ deck}}$ (MPa)	$= 300$	Maximum allowable tensile stress in deck
E_D	$= 243$	Number of elements of the deck in which stresses are checked
$d_{\min}$ (m)	$= 5$	Minimum distance between the anchor position of two consecutive cables

According to the design regulations, the vertical displacement on the deck must be evaluated in SLS, i.e., only in the intact model for the Load Cases 0 and 1 (eqns (4b) and (4c)). On the other hand, cables and deck stresses are evaluated in ULS for the intact model and in EELS for damaged configurations (eqns (4d), (4e), (4f)). It is important to note that for the intact model, it is not necessary to check the stress constraints in the Load Case 0, as these values are always more unfavourable in the Load Case 1. In addition, a minimum distance between two consecutive cables was imposed in order to reduce the chances of a vehicle hitting more than one cable (eqn (4g)). A total number of 21,779 design constraints were considered in the fail-safe optimization problem.

3.3 Numerical results

The layout of the initial design is presented in Fig. 4. The cables are equally spaced along each span and a cable area of 0.03 m² has been considered, with different prestressing forces.

Fig. 5 shows the final cable arrangement obtained and Fig. 6 shows the evolution of the objective function.

Figure 4: Cable arrangement of the initial design.

Figure 5: Optimum cable area distribution of the fail-safe optimization.

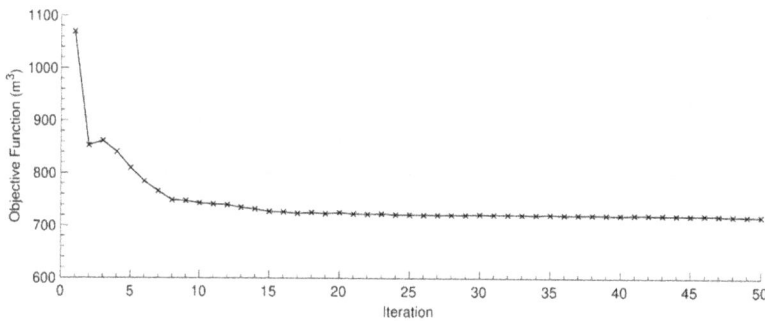

Figure 6: Evolution of the objective function in the fail-safe optimization.

The final steel volume of the cable system corresponds to 719.76 m³, this is, a penalty in steel volume of 13.50% with respect to the optimization of the intact model but this value is still lower than the volume of material of the real bridge. This leads to eight additional cables in the side span and six additional cables in the main span. The values of the design variables appears in Fig. 7. Grey color corresponds to the cables anchored at tower 1, whereas green color is associated with cables anchored at tower 2.

The area of the cables in the lateral spans can be divided into two distinct groups. Cables anchored between deck coordinates (50–120) m and cables located between deck coordinate 150 m and the first tower. In the first group, the cables have significant areas with values between 0.04 m² and 0.06 m². In addition, most of these cables are located around the intermediate pile of the lateral span. In the second group, the area of the cables is lower, with the aim of providing vertical support to the deck. In the main spans, there is a group of four cables of great length and similar areas around 0.04 m² which are anchored in the central tower and in the deck near the side towers. These cables control the horizontal displacement of the central tower tip. The remaining cables are arranged in a fan-shaped distribution as in conventional cable-stayed bridges. The areas of these cables grow from the towers towards the center of the span with areas between 0.02 m² and 0.055 m².

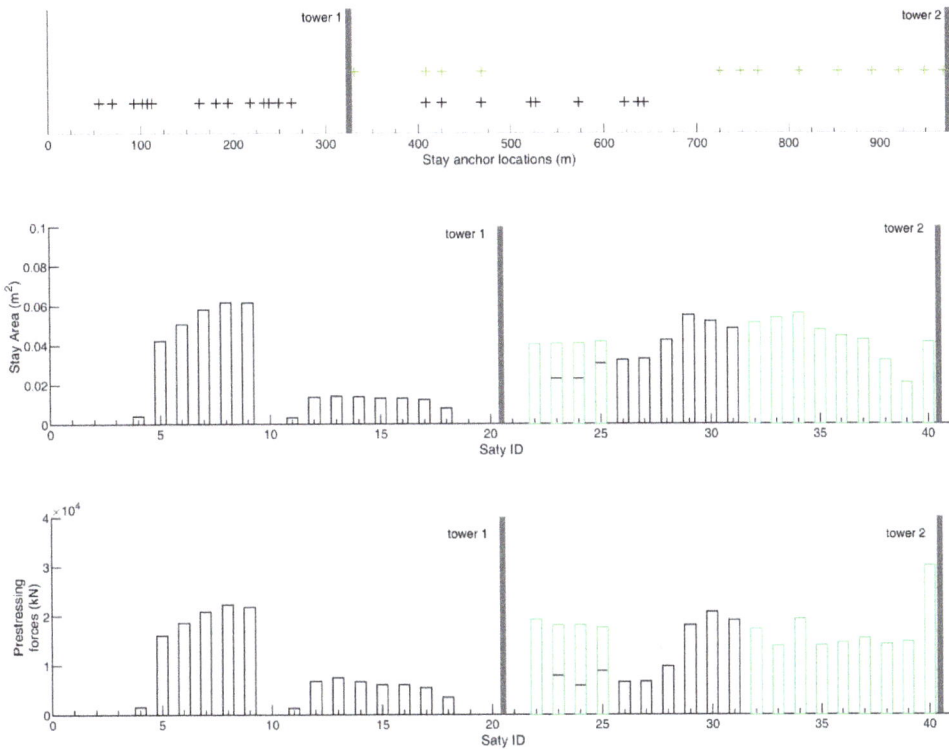

Figure 7: Values of design variables at the optimum design (only half a bridge is shown).

Regarding the active constraints, displacement constraints in the deck are active in the intact model for the Load Case 0 and 1, while displacement constraints of the tower head are active only in Load Case 1, as can be seen in Figs 8 and 9. As for the damaged models, it can be observed that most of the active stress constraints in cables occur in the vicinity of the damaged cable, in some cases activating up to eight cables simultaneously. Regarding the deck stress constraints in damaged models, they occur when cables break in the main span.

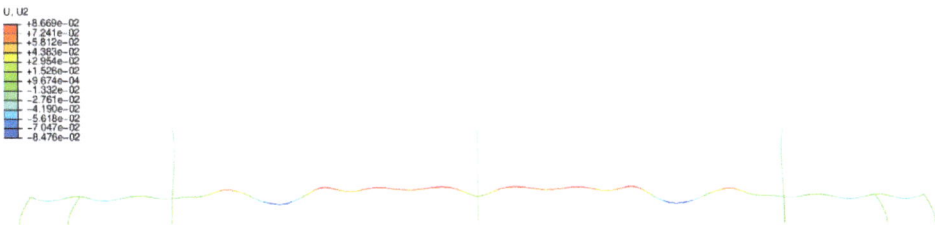

Figure 8: Displacements in the intact model (M_0) for the Load Case 0 (scale factor = 200).

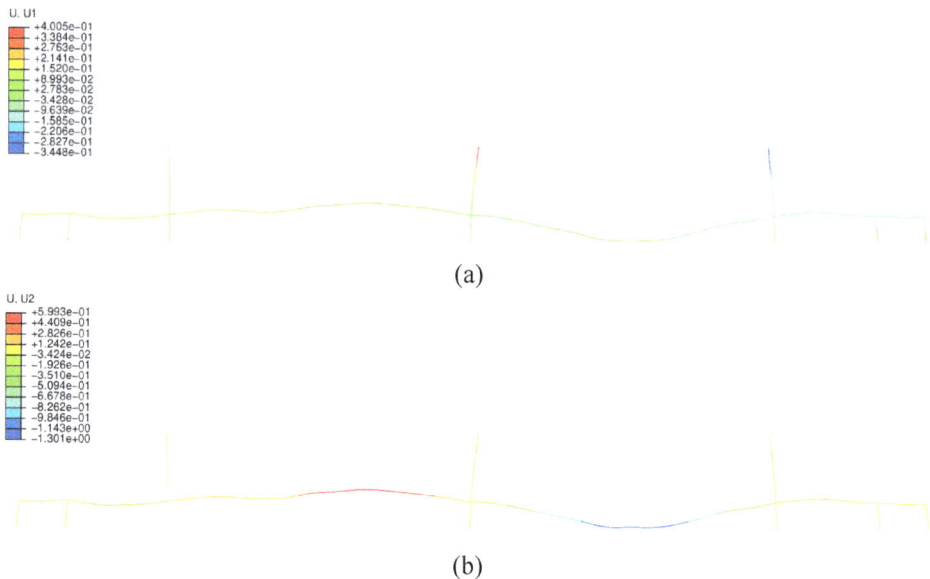

Figure 9:   Displacements in the intact model (M_0) for the Load Case 1. (a) Active horizontal displacement of the central tower head (scale factor = 40); and (b) Active vertical displacement of the deck (scale factor = 40).

4 CONCLUSIONS

Several conclusions can be drawn from this work:

1. Cable breakage has been successfully incorporated into the optimization of cable-stayed bridges, with satisfactory results.
2. Apart from the cable removal, the dynamic effect of the rupture on the remaining structure is taken into account by the application of two impact forces at the anchorage locations of the broken cable.
3. The consideration of a cable rupture into the optimization process greatly influences the volume of the optimum cable system, reaching a penalty volume of 13.5%.
4. There are active constraints in both the intact and damaged models, demonstrating the importance of the application of fail-safe optimization strategies, leading to a minimum steel volume increase.

ACKNOWLEDGEMENT

The research leading to these results has received funding from the Galician Government through research grant ED431C 2021/33.

REFERENCES

[1]   Bhatti, M., Raza, S.M. & Rajan, S.D., Preliminary optimal design of cable-stayed bridges. *Engineering Optimization*, **8**(4), pp. 265–289, 1985. DOI: 10.1080/03052158508902493.
[2]   Ohkubo, S. & Taniwaki, K., Shape and sizing optimization of cable-stayed bridges. *Optimization of Structural Systems and Industrial Applications*, pp. 529–540, 1991.

[3]   Simões, L.M.C. & Negrão, J.H.O., Sizing and geometry optimization of cable-stayed bridges. *Computers and Structures*, **52**(2), pp. 309–321, 1994. DOI: 10.1016/0045-7949(94) 90283-6.

[4]   Negrão, J.H.O. & Simões, L.M.C., Optimization of cable-stayed bridges with three-dimensional modelling. *Computers and Structures*, **64**(1), pp. 741–758, 1997. DOI: 10.1016/S0045-7949(96)00166-6.

[5]   Simões, L.M.C. & Negrão, J.H.O., Optimization of cable-stayed bridges with box-girder decks. *Advances in Engineering Software*, **31**(6), pp. 417–423, 2000. DOI: 10.1016/S0965-9978(00)00003-X.

[6]   Long, W., Troitsky, M.S. & Zielinski, Z.A., Optimum design of cable-stayed bridges. *Structural Engineering and Mechanics*, **7**(3), pp. 241–257, 1999. DOI: 10.12989/sem.1999.7.3.241.

[7]   Martins, M.B.A., Simões, L.M.C. & Negrão, J.H.J.O., Optimum design of concrete cable-stayed bridges. *Engineering Optimization*, **48**(5), pp. 772–791, 2016. DOI: 10.1080/0305215X.2015.1057057.

[8]   Montoya, M.C., Hernández, S. & F. Nieto, F., Shape optimization of streamlined decks of cable-stayed bridges considering aeroelastic and structural constraints. *Journal of Wind Engineering and Industrial Aerodynamics*, **177**, pp. 429–455, 2018. DOI: 10.1016/j.jweia.2017.12.018.

[9]   Sun, B., Zhang, L., Qin, Y. & Xiao, R., Economic performance of cable supported bridges. *Structural Engineering and Mechanics*, **59**(4), pp. 621–652, 2016. DOI: 10.12989/sem.2016.59.4.621.

[10]   Hassan, M.M., Nassef, A.O. & El Damatty, A.A., Determination of optimum post-tensioning cable forces of cable-stayed bridges. *Engineering Structures*, **44**, pp. 248–259. DOI: 10.1016/j.engstruct.2012.06.009.

[11]   Sung, Y.-C., Chang, D.-W. & Teo, E.-H., Optimum post-tensioning cable forces of Mau-Lo Hsi cable-stayed bridge. *Engineering Structures*, **28**(10), pp. 1407–1417, 2006. DOI: 10.1016/j.engstruct.2006.01.009.

[12]   Baldomir, A., Hernandez, S., Nieto, F. & Jurado, J.A., Cable optimization of a long span cable stayed bridge in La Coruña (Spain). *Advances in Engineering Software*, **41**(7), pp. 931–938, 2010. DOI: 10.1016/j.advengsoft. 2010.05.001.

[13]   Hassan, M.M., Optimization of stay cables in cable-stayed bridges using finite element, genetic algorithm, and B-spline combined technique. *Engineering Structures*, **49**, pp. 643–654, 2013. DOI: 10.1016/j.engstruct.2012.11.036.

[14]   Baldomir, A., Tembrás, E. & Hernández, S., Optimization of cable weight in multi-span cable-stayed bridges. Application to the Forth Replacement Crossing. *Proceedings of Multi-Span Large Bridges*, 2015.

[15]   Cid, C., Baldomir, A. & Hernández, S., Optimum crossing cable system in multi-span cable-stayed bridges. *Engineering Structures*, **160**, pp. 342–355, 2018. DOI: 10.1016/j.engstruct.2018.01.019.

[16]   Martins, A.M.B., Simões, L.M.C. & Negrão, J.H.J.O., Optimization of cable-stayed bridges: A literature survey. *Advances in Engineering Software*, **149**, 102829, 2020. DOI: 10.1016/j.advengsoft. 2020. 102829.

[17]   MATLAB r2016b Documentation.

[18]   ABAQUS 6.14.2 Documentation.

[19]   Wolff, M. & Starossek, U., Cable loss and progressive collapse in cable-stayed bridges. *Bridge Structures*, **5**(1), pp. 17–28, 2009. DOI: 10.1080/ 1573248090277 5615.

[20]   Post Tensioning Institute (PTI), Recommendations for stay-cable design, testing and installation, 2007.

[21] Baldomir, A., Hernández, S., Romera, L. & Díaz, J., Size optimization of shell structures considering several incomplete configurations. *53rd AIAA/ASME/ASCE/AHS/ASC Structures, Structural Dynamics and Materials Conference*, Honolulu, Hawaii, 2012. DOI: 10.2514/6.2012-1752.

[22] Carter, M., Kite, S., Hussain, N., Seywright, A., Glover, M. & Minto, B., Forth Replacement Crossing: Scheme design of the bridge. *IABSE Symposium Report*, **96**, pp. 107–116, 2009. DOI: 10.2749/222137809796088305.

30 YEARS' EXPERIENCE ON THE OPTIMIZATION OF CABLE-STAYED BRIDGES

ALBERTO M. B. MARTINS[1], LUÍS M. C. SIMÕES[1],
JOÃO H. J. O. NEGRÃO[2] & FERNANDO L. S. FERREIRA[1]
[1]ADAI, Department of Civil Engineering, University of Coimbra, Portugal
[2]Department of Civil Engineering, University of Coimbra, Portugal

ABSTRACT

Cable-stayed bridges optimization consists of finding the stiffness and mass arrangement of the load-bearing members (deck, towers and cable-stays) and the cable forces distribution, aiming to minimize cost and to achieve an adequate structural behaviour under static and dynamic loading. The first works on this topic were reported over 40 years ago but it is attracting growing interest with more than half of the publications in the last decade. The main goal of this paper is to share the perspective of this research groups 30 years' experience in this domain. This paper starts with an overview of the optimum design of cable-stayed bridges followed by a presentation of previous research works by the authors. Current research and future developments envisaged are also referred to. The first works consisted of the optimization of steel bridges considering three-dimensional modelling, box-girder decks, seismic action and uncertainty-based optimization. The optimum design of concrete bridges and the simultaneous optimization of structure and control devices in steel footbridges subjected to pedestrian-induced vibrations and steel bridges under seismic action followed. The optimization of bridges with complex geometries and other cable-supported concrete bridges, like extradosed bridges and under-deck cable-stayed bridges, are subjects of recent research. The optimization of long-span and multi-span bridges, including novel cable arrangements, and the optimum design considering robustness are of major relevance in future developments. Although with limited scope at present, given the problem size, it is expected an increasing use of metaheuristic algorithms, artificial neural networks and surrogate models.
Keywords: *cable-stayed bridges, optimization, cable forces, optimum design, sizing design variables, shape design variables.*

1 INTRODUCTION

Cable-stayed bridges are widely used all over the world for medium-to-long spans. Their popularity is due to its structural and construction efficiency but also to economic and aesthetic advantages, due to their elegant and transparent appearance.

Modern cable-stayed bridges feature multiple inclined cable-stays providing the deck with a continuous support and a natural prestressing. This allows spanning long distances with slender decks (main span-to-depth ratio of about 250). The cable-stays transfer the loads to the towers that, acting in compression, transmit these loads to the foundations. The cable forces distribution and the stiffness of the load-bearing members (deck, towers and cable-stays) determine the behaviour of these highly redundant structures. There is a wide range of structural solutions for cable-stayed bridges, especially, for small and medium spans. However, the symmetrical typology with three spans and two towers has been commonly adopted for medium-to-long spans. The towers are usually made of concrete, with the deck being of concrete, steel or steel–concrete composite. The historical background, a detailed review of the main features and a comprehensive analysis of the structural behaviour of cable-stayed bridges can be found in several references [1]–[5].

Cable-stayed bridges design is a complex problem that comprises defining the structural system, finding the members cross-sectional sizes, computing the cables forces distribution, geometrical non-linear effects and erection stages. The time-dependent behaviour of concrete and the dynamic actions add more complexity to this design problem. Therefore, optimization techniques are especially appropriate for solving this large and complex design problem seeking cost minimization and structurally efficient solutions considering both, static and dynamic actions.

A convex optimization problem features a convex objective function and a convex feasible domain. The feasible domain is convex if all the inequality constraints are concave and the equality constraints are linear. A convex optimization problem has only one minimum, and the Karush–Kuhn–Tucker (KKT) conditions are sufficient to establish it. The domain in structural optimization is in general nonconvex. However, convex optimization is relevant because a sequence of approximate problems is used in explicit optimization [6].

Convex optimization and nonconvex optimization strategies can be used to iteratively modify the design variables seeking a design improvement. Convex optimization strategies need the derivatives of the objective function and all the design constraints with respect to the design variables. This information (sensitivity analysis) is used to define the search direction along which the current design variables are modified seeking an optimum solution. These approaches converge in polynomial time to a local (not necessarily global) optimum solution. Nonconvex optimization strategies do not use sensitivity analysis and consist of different procedures, such as random search, branch-and-bound or metaheuristic algorithms (genetic algorithms, evolutionary algorithms, particle swarm optimization, simulated annealing). Although they are easier to implement, they feature an exponential convergence time with respect to the number of design variables to find the optimum solution. They usually finish with the best solution found so far, not even guaranteeing to be a local optimum unless the solution satisfies the KKT optimality conditions.

The optimization problem of cable-stayed bridges may be nonconvex and the feasible domain may be non-connected when considering dynamic loading. The cable-stayed bridge optimization usually features a large number of design variables and design constraints, which are nonlinear and conflicting. This leads to a complex design space and a computationally costly problem. Considering this, an efficient convex optimization technique associated with multiple starting points was adopted preferably to a metaheuristic optimization technique. Finding the active set of constraints in gradient-based nonlinear programming algorithms may pose a considerable difficulty in solving problems with hundreds or thousands of constraints. A procedure was envisaged considering all constraints according to their relative merit. There are several approaches for this constraint aggregation, such as, the Kreisselmeier–Steinhauser function [7] or the least p-norm [8].

In previous works, the optimization of cable-stayed bridges was formulated as a multi-objective problem which is solved by the minimization of a convex scalar function obtained via an entropy-based approach. This function aggregates all the design objectives and creates a convex approximation close to the boundaries of the original nonconvex domain. This approach avoids the complicated procedure of finding the active set of constraints in a problem with a significant number of constraints. All the constraints are included in the scalar function with different probabilities of becoming active. As iterations proceed, there is decreasing uncertainty about which of the constraints are more important to find the optimum. This procedure reduces the cost objective by improving its value with respect to the previous iterations and simultaneously keeps all the constraints within limits. The minimization of the scalar function is combined with a multi-start strategy to obtain optimal local solutions and the best is selected as the optimum design.

A recent literature survey by Martins et al. [9], comprising a detailed review of 90 articles, revealed that the first works on this topic were presented over 40 years ago. However, it can be stated that represents a topic of growing interest with more than half of the articles published in the last decade. With 30 years' experience and 26 articles published and indexed in the Web of Science and Scopus electronic databases, this paper aims to share the perspective of our research group. Future developments are expected, therefore, it also aims at contributing to draw the attention of future researchers in this topic. These publications were analysed and a detailed review was conducted to identify the main characteristics, the principal conclusions and contributions of each.

In Section 2, the formulation of the optimum design problem of cable-stayed bridges is described. Section 3 presents a detailed review of previous articles by the authors. The most recent works are also pointed out. Finally, Section 4 presents some concluding remarks, the current trends and possible future developments within this research topic.

2 OPTIMUM DESIGN OF CABLE-STAYED BRIDGES

2.1 Overview

In the optimum design of cable-stayed bridges, the objective function is usually formulated based on criteria related to structural efficiency and/or economy, aiming to minimize total cost, total strain energy or some norm of the deck vertical displacements and towers horizontal displacements. Regardless of the objective function chosen, safety and service criteria should be met to achieve a feasible design. The problem can be posed as a single objective optimization to minimize the cost of the structure while satisfying displacement and stress constraints imposed according to the design codes.

In previous works by the authors, the optimum design problem of cable-stayed bridges is formulated as a multi-objective optimization problem from which a Pareto optimal solution vector is found. This means that no other feasible vector exists that could decrease one objective without increasing at least another one.

2.2 Structural analysis

The finite element method was used for assessing the structural response under static and dynamic loadings. A linear elastic analysis is commonly adopted and various loading cases should be considered. These should refer to the complete bridge under permanent load and live load placed to obtain the most unfavourable effects.

The balanced cantilever method is usually considered the reference erection method for these bridges. In this method, the geometry of the structure, the displacements and stresses change during the construction process. Therefore, the analysis should consider the construction stages aiming to evaluate the corresponding displacements and stresses that should be included as constraints in the optimization problem.

In these bridges, there are three main sources of geometric nonlinearities: the nonlinear axial force–elongation relationship for the inclined cable stays due to the sag caused by their own weight; the nonlinear axial force and bending moment–deformation relationships for the towers and the deck under combined bending and axial forces; and the geometry change caused by large displacements. A second-order elastic analysis can be adopted to account for the geometric nonlinear effects. To consider the cable sag, the cables can be modelled as truss elements with an equivalent modulus of elasticity given by Ernst formulation. Two approaches were used to include the geometrical nonlinearities in the deck and towers, namely, the equivalent lateral force method and computing the stiffness matrix of the elements in the deck and towers with the elastic ($\underline{K}_E$) and geometric ($\underline{K}_G$) contributions. Step-by-step integration of the dynamic equilibrium equation and modal/spectral approaches were utilized to access the structural response under dynamic actions.

2.3 Design variables

The design variables, $\underline{x}$, can be grouped in three sets: sizing, shape and mechanical. The first set concerns the cross-sectional dimensions of towers, deck and cable-stays. The second set pertains the bridge geometry (tower height, lateral and central span lengths, cable anchor positions). Both sets directly influence the mass, stiffness and cost of the structure. The third set represents the cable-stays prestressing forces, which do not have direct impact in cost, but play a fundamental role in controlling the behaviour of cable-stayed bridges.

The cables forces optimization problem usually features one or two design variables per each cable-stay depending on if the cable force, the cable area or both are considered as design variables. Consequently, in modern cable-stayed bridges featuring a large number of cable-stays, this problem easily presents more than 20 design variables. An extra 10 to 20 design variables are added by considering sizing and shape design variables. This issue may justify the relatively small numbers of publications (18.4% of the cable forces optimization works and 14.7% of the optimum design works) using metaheuristic algorithms [9]. Some strategies were proposed to reduce the number of design variables by variable linking thus reducing the computational effort.

2.4 Design objectives

The first design objective concerns the cost minimization and can be expressed as

$$g_1(\underline{x}) = \frac{C}{C_0} - 1 \le 0 , \tag{1}$$

where C is the current cost of the structure and C_0 is a reference cost, that represents the initial cost of each analysis and optimization cycle. This approach makes the cost always one of the main objectives for the optimization algorithm.

A second set of objectives refer to limiting the deck vertical displacements and the towers horizontal displacements, aiming at the desired deck profile at the end of construction and the minimization of the tower bending deflections

$$g_2(\underline{x}) = \frac{|\delta|}{\delta_0} - 1 \le 0 , \tag{2}$$

where δ and δ_0 are the displacement value and the limit value for the displacement under control, respectively.

The stress objectives for the deck and towers members are defined based on the provisions of the relevant design codes according to the structural material used. In general, these goals can be expressed by

$$g_3(\underline{x}) = \frac{\sigma}{\sigma_{allow}} - 1 \le 0 , \tag{3}$$

where σ and σ_{allow} are the acting stress and the corresponding allowable stress, respectively. For concrete members, different values of the allowable stress should be considered for service conditions and for strength verification. For service conditions, appropriate allowable values should be considered according to the different concrete strength in tension and in compression. For strength verification, the allowable value should represent the structural concrete member's resistance, including reinforcement, evaluated according to acting internal forces, such as, bending and axial force, shear force or biaxial bending and axial force.

Another set of objectives refers to the stresses in the cable-stays which can be written as

$$g_4(\underline{x}) = \frac{\sigma}{k \cdot f_{pk}} - 1 \le 0, \tag{4}$$

$$g_5(\underline{x}) = 1 - \frac{\sigma}{0.1 f_{pk}} \le 0, \tag{5}$$

where σ and f_{pk} are the acting stress in the stays and the characteristic value of the prestressing steel tensile strength, respectively. The value of k in eqn (4) is usually equal to 0.55 during construction, 0.45 or 0.50 for service conditions and 0.74 for strength verification. Eqn (5) concerns to a lower limit for tension in the stays to ensure their structural efficiency.

More than 1,000 design goals can be easily expected when solving this optimization problem due to the structural discretization, the number of loadings and erection stages that need to be considered.

2.5 Objective function

A multi-objective optimization problem aims to minimize the set of all design objectives over the design variables. This can be expressed as a *minimax* problem, which is discontinuous and non-differentiable and consequently it is difficult to solve numerically. By using the Shannon/Jaynes Maximum Entropy Principle and Cauchy's arithmetic–geometric mean inequality [10] this problem is solved by replacing the *minimax* problem into the minimization of an unconstrained convex scalar function given by

$$\min F(\underline{x}) = \min \frac{1}{\rho} \ln \left[\sum_{j=1}^{M} e^{\rho(g_j(\underline{x}))} \right], \tag{6}$$

where $\underline{x}$ is the vector of N design variables, M is the number of design objectives and ρ is a control parameter which must not be decreased through the analysis and optimization process. This function is both continuous and differentiable and, thus, considerably easy to solve. This entropy-based optimization approach creates an inside convex approximation of the original nonconvex domain. The convex approximation parameter ρ is increased throughout the iterative procedure to make the convex underestimates closer to the original nonconvex domain.

Generally, in multi-objective optimization problems the designer assigns weighting factors to each goal according to its relevance. In this formulation, an entropy allocated factor is assigned to each goal, being modified in each iteration. The unbiased entropy-based factor assigned to each goal is the exponential of the goal (modified by the changes in the design variables) multiplied by the parameter ρ.

This algorithm requires a sequence of positive values of ρ increasing towards infinity. The aggregation parameter, ρ, is a user-defined parameter and can be computed by different schemes, such as, adaptive approach [11], [12] or the value which makes the objective function, eqn (6), stationary with respect to ρ [10]. The use of a constant parameter is also possible, but a value not large enough may lead to a conservative design. From a practical viewpoint values in the range 100–2,000 lead to the same results.

The design objectives, $g_j(\underline{x})$, do not have an explicit algebraic form, therefore, the problem is solved by an explicit approximation given by taking Taylor series expansion of all the objectives, around the current design variable vector, truncated after the linear term.

2.6 Sensitivity analysis

The sensitivity analysis is a fundamental aspect when using gradient-based optimization algorithms from which depends on the evolution of the analysis and optimization process and its accuracy. The sensitivity analysis provides the gradients of the objective function and all the design goals with respect to the design variables. From the various methods to perform the sensitivity analysis, the discrete direct method was used with both, analytical and semi-analytical derivatives. This is justified by the computational efficiency, the availability of the source code, the accuracy and that the number of design goals is far larger than the number of design variables.

The sensitivities of displacements with respect to the design variables are obtained by differentiating the static equilibrium equation. The stress sensitivities are calculated from the chain derivation of the finite element stress–displacement relation.

The nonlinear structural response (including geometrical nonlinearities and dynamic actions) to changes in the design variables is approximated by the first-order Taylor series approximation. To ensure the accuracy of this linear approximation in each analysis-and-optimization cycle, bound constraints with move limits should be imposed on the variation of the current value of each design variable. The values of the allowable variations can be selected depending on the nonlinearity of the goal functions.

3 PREVIOUS AND CURRENT RESEARCH

3.1 Optimum design of steel bridges

Our first work on the optimization of steel bridges cast as a multi-objective problem with goals of minimum cost and stresses [13] used the two-dimensional analysis of a three-span symmetrical bridge including erection stages. The cable anchor positions on the deck and towers and the cross-sectional sizes of the structural members were considered as design variables. The results emphasised the relevance of considering the erection stresses and the cable anchor positions as design variables aiming at cost reduction.

Aiming at a more computationally efficient analysis-and-optimization procedure, Negrão and Simões [14], [15] focused on the development and implementation of an analytical sensitivity analysis and optimization for cable-stayed bridges design. The optimum design was posed as a multi-objective optimization with goals of minimum cost of material, stresses and displacements. Cable anchor positions on the deck and towers and cross-sectional sizes of the structural members were considered as design variables.

The optimization of steel bridges considering box-girder followed [16]–[18]. This solution for the deck is very effective for long span bridges, due to its high torsional stiffness and streamlined profile, favouring a good aerodynamic behaviour. Additional design variables were considered, namely, shape and sizing design variables of the box-girder, cross-sectional sizes of cable-stays and towers, cables prestressing forces and shape design variables (cables anchor positions in the deck and towers, towers height, lateral and central span lengths). The multi-criteria approach for the optimum design problem considered goals of minimum cost, maximum stresses, minimum stresses in stays and deflections under dead load condition.

The seismic action is of major concern when designing cable-stayed bridges built on earthquake prone areas. Thus, the optimum design of steel bridges under seismic action was also investigated [19] considering both, modal response spectrum and time-history approaches. The analytical sensitivity analysis was carried out for both approaches providing the structural response to seismic action with respect to changes in the design variables. The cables areas and the cross-sectional dimensions of the deck and towers were considered as design variables. It is worth referring that both approaches provided adequate solutions for optimization of steel bridges under seismic vibrations. However, each method poses specific problems associated with either the code implementation or required runtime. Generally, the modal/spectral approach should be preferred, but the basic assumptions of this method make it inadequate for non-linear problems. Given the high computational cost of the time-history approach, in an analysis-and-optimization procedure, it should be reserved for non-linear problems.

The optimum design of steel bridges subjected to seismic action and the effect of control devices [20] started a number of works on this subject. A time-history approach was adopted for accessing the structural response under different earthquake recordings. The multi-objective optimization considered design goals of cost, and different evaluation criteria related to internal forces and displacements in selected locations. A total of 37 design variables comprising shape, sizing and control were considered. The results shown that the integrated structure-and-control optimization provided minimum cost solutions with improved dynamic performance.

The simultaneous structural-control optimization problem of steel bridges under seismic action was addressed considering three-dimensional modelling and erection stages [21]. The modal/spectral approach was used to evaluate the structural response under seismic action defined according to Eurocode 8 provisions. More than 50 design variables were considered, namely, cross-sectional sizes of deck and towers, cables cross-sectional areas and prestressing forces, bridge geometry (cables anchor positions, spans lengths tower geometry), and stiffness and damping of the deck–tower connection. Different numbers of cables were considered in the multi-start strategy combined with the convex optimization algorithm. A numerical example with three cases concerning the deck-tower connection was analysed: free, fixed and with viscous dampers. The later presented the most cost-effective solution.

3.2 Optimum design of concrete bridges

The calculation of the cable forces distribution is a unique feature of cable-stayed bridges. These forces are crucial to control the construction process and, thus, achieving the desired geometry and stress distribution for the complete bridge. Many researchers focused on the cable forces optimization problem. The time-dependent behaviour of concrete adds difficulty to this problem which was specifically addressed in a previous work [22]. The problem was posed as a multi-objective optimization with goals of stresses, and minimum deck vertical displacements and minimum towers horizontal displacements. To adequately evaluate the bridge behaviour during construction and at completion, the structural analysis included the construction sequence, the load history and the time-dependent effects of creep, shrinkage and ageing of concrete. The time-dependent effects were evaluated according to Eurocode 2 formulations. The creep model was based on linear viscoelasticity and takes into account ageing effects. Shrinkage strains are time-dependent but stress independent. The creep function was approximated by a Dirichlet series to evaluate the concrete creep deformations under variables stress. The cable-sag effect was considered by modelling the cables as bar elements with equivalent modulus of elasticity given by Ernst formulation. A numerical example with a total of 32 design variables corresponding to two sets (installation and adjustment) forces was considered. A computer program developed in MATLAB environment was used to perform the structural analysis, sensitivity analysis and optimization. The second-order effects in the deck and towers were latter included [23], [24] through a second order elastic analysis by the equivalent lateral force method. The results shown the relevance of considering the construction stages, the time-dependent effects and the geometrical nonlinearities for adequately predict the bridge behaviour and compute the cable prestressing forces.

The construction stages, the concrete time-dependent effects and the geometrical nonlinearities were also considered in the optimum design of concrete bridges with different deck cross-sections [25] and including deck prestressing [26]. The optimum design was formulated as a multi-criteria problem with objectives of minimum cost, minimum displacements, stresses under service conditions and strength criteria. Three options were analysed for the deck cross-section: beam-and-slab, single-cell box and tri-cell box. Rectangular hollow sections were considered for the towers. A total of 68 design variables were considered corresponding to the cables areas and prestressing forces, and the cross-sectional sizes of the deck and towers. In the optimum design of concrete bridges, special attention should be paid when computing the concrete members' resistance that should include the steel reinforcement. Moreover, an additional difficulty arises when computing the sensitivities of the strength design goals due to the fact that the resistance of each concrete member depends on the correspondent cross-sectional design variables. This was solved by calculating the sensitivities of these design objectives in normalized form.

More recently, the optimization of concrete bridges under seismic action was studied [27]. The computer program previously developed in MATLAB environment was improved to allow three-dimensional modelling, static and dynamic analysis. The finite element method was used for the structural analysis considering static loading (dead load and road traffic live load), the time-dependent effects and the geometrical nonlinearities. The modal superposition method was used to access the structural response under seismic action defined according to Eurocode 8 elastic response spectrum. The complete quadratic combination (CQC) was used for modal combination due to the modal coupling that is present in the dynamic response of these bridges. The multi-criteria optimum design considered design objectives of cost, deflections, natural frequencies and stresses for both, serviceability and ultimate limit states defined according to Eurocode 2 provisions. The design variables were the cable-stays areas and prestressing forces, the deck and towers cross-sections. The modal analysis requires obtaining the stiffness matrix in the dead-load permanent state. Thus, the second-order-effects in the deck and towers were considered by computing the stiffness matrix with the elastic and geometric contributions. The seismic action governs the design of these structures, being especially demanding for the strength verification of the structural members, mainly the towers. The optimum seismic design of concrete cable-stayed bridges should be further investigated with a special focus on the geometry and seismic design of towers and the arrangement of the cable suspension system. Moreover, the spatial variability of the seismic ground motion, the soil–structure interaction, the use of passive and active control devices and the simultaneous optimization of the structure and control devices should be considered in future developments.

3.3 Optimum design of steel footbridges

This topic has been a subject of major research with a focus on the simultaneous optimization of structure and control devices. The first work on this subject was published in 2010 [28] consists of a two-dimensional model of a two span cable-stayed footbridge subjected to a crowd of joggers during a running event and using one active tendon.

The least cost solution of a passive and active non-symmetric cable-stayed footbridge was considered in Ferreira and Simões [29]. The active bridge was controlled using a single active tendon and for the passive bridge, a passive damper was used in the vertical deck–tower connection. The finite element method was used for the two-dimensional analysis of the bridge under both, static and dynamic loadings. A time-history procedure was used to evaluate the displacements and accelerations of the bridge subjected to a running event. A total of 29 design variables were considered, corresponding to shape, sizing, mechanical and control design parameters. The multi-objective optimization reduced simultaneously cost, accelerations and displacements while satisfying allowable stresses. Numerical results shown that both passive and active optimum designs are efficient, with different geometry, mass distribution and cost (22% higher in the passive design).

Further research comprised the optimum design of a two-span non-symmetric cable-stayed footbridge, subjected to a running event and considering four different control techniques [30], namely, no control, viscous dampers (VD), and VD with passive or semi-active tuned-mass dampers (TMDs). A total of 27 design variables were considered in the multi-objective optimization including static and dynamic design goals. Static and dynamic design criteria were considered in the multi-objective optimization problem. The results revealed that the solution with VD is the most cost effective, because the damper adds an important control action to the most relevant modes. This was achieved due to the simultaneous optimization of tower position. Furthermore, the optimum structural solution (cross-sectional sizes and geometry) for the different cases is rather different depending on the control devices, meaning that the integrated structural-and-control optimization is a design advantage.

The computational model was further improved for the three-dimensional optimum design of steel footbridges [31] with a focus on the control of the synchronous lateral excitation, "lock-in", that occurs in long-span footbridges. The optimum design was formulated as a multi-objective problem with goals of minimum cost, static and dynamic design criteria. A total of 43 sizing, geometry and control design variables were considered. VD were adopted as control devices. An analytical sensitivity analysis was used, including a formulation for finding the pedestrian "lock-in" sensitivities. The algorithm provides minimum cost solutions which simultaneously satisfy the ultimate and service limit considered, such as, stresses throughout the structure, buckling of the structure and the members, displacements, accelerations and the dynamic stability of the structure when subject to synchronous lateral excitation. The algorithm was able to shift the natural frequencies to fall outside the critical ranges for the vertical acceleration and for the "lock-in" phenomena. To satisfy the vertical acceleration and the "lock-in" criteria in the remaining modes, the algorithm changed the tower position and the cables anchor positions, and modified the stiffness and damping properties of the deck-tower connection. These properties are important because if a stiff tower-deck connection is used the structural frequencies will increase, but no damping will be added. On the other hand, with a free connection the frequencies will decrease but, also, no damping will be added. The connection properties are also important for the static design criteria, in particular to control the deck displacements.

3.4 Optimum design under uncertainty

This topic was firstly addressed considering the Two-Phase Method for fuzzy optimization of steel cable-stayed bridges [32]. This method is based on the Fuzzy Set Theory and corresponds to a non-probabilistic description of uncertainty. In the first phase, the fuzzy solution is obtained by using the Level Cuts Method and in the second phase the crisp solution, which maximizes the membership function of fuzzy decision-making, is found by using the Bound Search Method. Fuzzy goals were generated involving Young's modulus, stress limits and loading. Shape and sizing design variables were considered and their values were assumed to be deterministic. The optimization problem was posed as the minimization of bridge cost, stresses and displacements. This approach provided solutions featuring high design levels and materials savings when compared to the deterministic design.

The reliability-based optimum design of steel cable-stayed bridges [33] and glulam cable-stayed footbridges [34] was also studied. The first-order second moment method (FORM) was used to compute the reliability indices associated with various limit state functions. Second-order bounds of the probability of failure were considered to evaluate the structural system probability of failure. The bound intervals obtained show that most of the nearly 1000 limit states considered are highly correlated. A discrete reliability analytical sensitivity analysis was derived and used in the optimization algorithm. The multi-objective optimum design problem was formulated as the minimisation of stresses, displacements, reliability and bridge cost. Shape, sizing and mechanical design variables were considered. Material properties and loadings were considered as random variables. Optimum solutions featuring cost reduction and reduced probability of failure were obtained. Although, in these examples, the probability of failure refers to critical stresses throughout the structure, induced by loadings, other failure modes or criteria could be used as well, such as the excessive deflection or cable under-stressing. The advanced simulation method combined with the response surface method will be proposed in future research.

3.5 Recent research

The optimization of other types of concrete cable-supported bridges, such as, extradosed bridges [35] and under-deck cable-stayed bridges [36] was a focus of recent research. The computational tool previously developed for the optimization of concrete cable-stayed bridges was generalized to solve these problems. The optimum design of both structures aims to find an adequate balance between the stiffness of the main girder and the cables suspension effect, depending on the cross-sectional dimensions and the prestressing forces. Besides the cross-sectional areas and prestressing forces of the extradosed or under-deck cables, these problems require considering the cross-sectional areas and prestressing forces of the internal tendons in the deck. The time-dependent effects of concrete and prestressing steel should be included. Numerical examples comprising a symmetrical extradosed concrete bridge with a total length of 330 m and a main span of 150 m were analysed. Static loading (dead load and road traffic live load) and seismic action were considered in the optimum design considering 40 design variables and more than 1,600 design objectives. For these examples, the seismic action governs the design when medium or high intensity seismic action is considered. There is a wide range of structural solutions for these bridge types, therefore, requiring additional research. Tower geometry, towers–deck–piers connection, deck cross-section design and the number of cables influence the structural behaviour of extradosed bridges. The use of passive and active control devices and the simultaneous optimization of the structure and control devices should be considered in future developments. Further studies concerning under-deck cable-stayed bridges and the optimization of tied-arch bridges are expected in future developments.

The optimum design of bridges with complex geometries was another topic of current research. The computer program previously used for the three-dimensional optimum design of steel footbridges [31] was improved for the optimum design of curved bridges with VD as control devices [37]. Numerical examples comprising different bridge lengths (180, 220, 260, 300 and 340 m) were analysed. A total of 59 design variables were considered, including, bridge geometry (tower shape, number of cables and their location), cross-sectional sizes, properties of control devices and cables prestressing. The optimization problem was formulated with objectives of least bridge cost, static and dynamic design criteria. From the results it could be stated that the design is governed by dynamic comfort requirements in particular the horizontal and vertical accelerations and the synchronous lateral instability, "lock-in". The algorithm was able to control the relevant modes, either by increasing/decreasing the frequencies to fall outside the critical range or adding damping to meet the dynamic design criteria. The tower location (main span-to-total length ratio) is one of the design variables that substantially influences the static response and the dynamic control of both, horizontal and vertical modes of vibration. The optimum solutions feature "A"-shaped towers with fan cable layout. This is relevant in curved cable-stayed bridges to reduce the tower torsion due to the different cable forces in both sides of the tower. The results revealed that the static and dynamic design coupling becomes more important for curved bridges design as the cable tensioning problem becomes more complex and the vibration modes exhibit simultaneous vertical, horizontal and torsional responses.

The algorithm was further improved for studying cable-stayed footbridges with "S"-shaped deck and external pylons [38]. A special attention was paid to generating initial feasible designs to start the optimization process. Given that dampers cannot be located at the pylon–deck connection because it does not exist, the structure was controlled by passive and semi-active tuned mass dampers. A time-history analysis was adopted to consider the non-linear dynamic behaviour of the semi-active device. Two optimal solutions were sought using semi-active (S1) and passive (S2) dampers. The results shown that both solutions feature different geometries and cross-sectional dimensions. The pylon height is markedly higher for structure S2 because this improves the bridge dynamic behaviour. The results also revealed that the pylon acts like a dynamic leverage arm magnifying the vertical response of the deck. As previously stated, the simultaneous structural-and-control optimization provides superior results than optimizing the structure and control devices separately.

The most recent article concerns the optimum seismic design of a 350 m curved steel road bridge [39]. A time-history approach was used to access the dynamic response considering three seismic events. The spatial variability of the seismic ground motion and erection stages were considered. The deck–tower connection is a key aspect for the seismic response. The longitudinal displacement between deck and tower was released, the stiffness and damping of the transverse and vertical supports were considered as design variables. Linear viscous dampers were adopted as control devices. This dynamic problem is highly nonlinear, therefore, the design space features multiple non-connected domains. Two cases were considered: seismic action scaled to 30% (Case 1) and full seismic intensity (Case 2). The results revealed that the optimum design of Case 2 features 21.5% more cost than the optimum design of Case 1 and 43% more cost than a similar straight bridge. Moreover, the complexity of the curved bridge makes more difficult to find feasible starting solutions.

4 CONCLUSIONS AND FUTURE DEVELOPMENTS

The optimization of long-span bridges and multi-span bridges with novel cable arrangements, such as, crossing-cables [40] are drawing the attention of some researchers. The cable forces optimization problem is still a subject of interest for several researchers which are presenting novel approaches for solving this problem, mainly using metaheuristic algorithms [41], [42]. Therefore, new developments are expected in these topics.

Recently, it can be noticed an increasing use of metaheuristic algorithms, artificial neural networks and surrogate models [43] in the optimization field. The application of these techniques in the optimum design of cable-stayed bridges will be the focus of upcoming research.

The rapid increase in the computational resources available will contribute to the application of soft computing strategies in this domain. These are easier to implement and do not require a deep understanding of optimization concepts and sensitivity analysis techniques. Moreover, the increasing computational capacity will allow including more demanding problems, such as, response to wind and earthquakes, in the optimum design of these bridges. The shape and sizing optimization of bridge decks considering wind effects has been a subject of major research in recent years [44], [45]. This is a complex and relevant topic in the design of long-span bridges and, thus, further developments are expected.

The dynamic behaviour is of utmost importance in the design of these structures and the use of control devices plays a key role improving the structural response under dynamic actions. Therefore, the simultaneous optimization of structure and control devices will continue as a subject of major relevance in forthcoming research. The reliability-based optimum design and the robust design including, for example, cable losses scenarios, were not sufficiently addressed in previous research and will represent topics of interest for future developments.

REFERENCES

[1]   Troitsky, M.S., *Cable-Stayed Bridges: Theory and Design*, Crosby Lockwood Staples: London, 1977.

[2]   Podolny, W. & Scalzi, J.B., *Construction and Design of Cable-STAYED bridges*, 2nd ed., Wiley: New York, 1986.

[3]   Walther, R. (ed.), *Cable Stayed Bridges*, 2nd ed., Telford: London, 1999.

[4]   Gimsing, N.J. & Georgakis, C.T., *Cable Supported Bridges: Concept and Design*, 3rd ed., Wiley: Hoboken, NJ, 2012.

[5]   Svensson, H., *Cable-Stayed Bridges: 40 Years Of experience Worldwide*, 1st ed., Ernst, Wiley-Blackwell: Berlin, 2012.

[6]   Haftka, R.T. & Gürdal, Z., *Elements of Structural Optimization*, 3rd revised and expanded ed., Kluwer Academic Publishers, 1992.

[7]   Kreisselmeier, G. & Steinhauser, R., Systematic control design by optimizing a vector performance index. *IFAC Proceedings Volumes*, **12**(7), pp. 113–117, 1979.

[8]   Charalambous, C., Nonlinear least pth optimization and nonlinear programming. *Mathematical Programming*, **12**(1), pp. 195–225, 1977.

[9]   Martins, A.M.B., Simões, L.M.C. & Negrão, J.H.J.O., Optimization of cable-stayed bridges: A literature survey. *Advances in Engineering Software*, **149**, 102829, 2020.

[10]   Simões, L.M.C. & Templeman, A.B., Entropy-based synthesis of pretensioned cable net structures. *Engineering Optimization*, **15**(2), pp. 121–140, 1989.

[11]   Poon, N.M.K. & Martins, J.R.R.A., An adaptive approach to constraint aggregation using adjoint sensitivity analysis. *Struct. Multidiscipl. Optim.*, **34**(1), pp. 61–73, 2007.

[12]   Zhang, K.-S., Han, Z.-H., Gao, Z.-J. & Wang, Y., Constraint aggregation for large number of constraints in wing surrogate-based optimization. *Struct. Multidiscipl. Optim.*, **59**(2), pp. 421–438, 2019.

[13]   Simões, L.M.C. & Negrão, J.H.O., Sizing and geometry optimization of cable-stayed bridges. *Computers and Structures*, **52**(2), pp. 309–321, 1994.

[14] Negrão, J.H.O. & Simões, L.M.C., Three dimensional nonlinear optimization of cable-stayed bridges. *Advances in Structural Optimization*, eds B.H.V. Topping & M. Papadrakakis, Civil-Comp Press: Edinburgh, UK, pp. 203–213, 1994.

[15] Negrão, J.H.O. & Simões, L.M.C., Optimization of cable-stayed bridges with three-dimensional modelling. *Computers and Structures*, **64**(1–4), pp. 741–758, 1997.

[16] Simões, L.M.C. & Negrão, J.H.O., Optimization of cable-stayed bridges with box-girder decks. *Computer Aided Optimum Design of Structures V*, eds S. Hernandez & C.A. Brebbia, Wessex Institute of Technology: Southampton, UK, pp. 21–32, 1997.

[17] Negrão, J. & Simões, L, Shape and sizing optimization of box-girder decks of cable-stayed bridges. *Computer Aided Optimum Design of Structures VI*, eds S. Hernandez, A.J. Kassab & C.A. Brebbia, Wessex Institute of Technology: Southampton, UK, pp. 323–332, 1999.

[18] Simões, L.M. & Negrão, J.H.J., Optimization of cable-stayed bridges with box-girder decks. *Advances in Engineering Software*, **31**(6), pp. 417–423, 2000.

[19] Simões, L.M.C. & Negrão, J.H.J.O., Optimization of cable-stayed bridges subjected to earthquakes with non-linear behaviour. *Engineering Optimization*, **31**(4), pp. 457–478, 1999.

[20] Ferreira, F.L.S. & Simões, L.M.C., Optimum design of a controlled cable stayed bridge subject to earthquakes. *Structural and Multidisciplinary Optimization*, **44**(4), pp. 517–528, 2011.

[21] Ferreira, F. & Simões, L., Synthesis of three dimensional controlled cable-stayed bridges subject to seismic loading. *Computers and Structures*, **226**, 2020.

[22] Martins, A.M.B., Simões, L.M.C. & Negrão, J.H.J.O., Cable stretching force optimization of concrete cable-stayed bridges including construction stages and time-dependent effects. *Structural and Multidisciplinary Optimization*, **51**(3), pp. 757–772, 2015.

[23] Martins, A.M.B., Simões, L.M.C. & Negrão, J.H.J.O., Optimization of cable forces for concrete cable-stayed bridges. *Proceedings of the 14th International Conference on Civil and Structural and Environmental Engineering Computing*, eds B.H.V. Topping & P. Iványi, Civil-Comp Press: Stirlingshire, UK, Paper 227, 2013.

[24] Martins, A.M.B., Simões, L.M.C. & Negrão, J.H.J.O., Optimization of cable forces on concrete cable-stayed bridges including geometrical nonlinearities. *Computers and Structures*, **155**, pp. 18–27, 2015.

[25] Martins, A.M.B., Simões, L.M.C. & Negrão, J.H.J.O., Optimum design of concrete cable-stayed bridges. *Engineering Optimization*, **48**(5), pp. 772–791, 2016.

[26] Martins, A.M.B., Simões, L.M.C. & Negrão, J.H.J.O., Optimum design of concrete cable-stayed bridges with prestressed decks. *International Journal for Computational Methods in Engineering Science and Mechanics*, **17**(5–6), pp. 339–349, 2016.

[27] Martins, A.M.B., Simões, L.M.C. & Negrão, J.H.J.O., Optimization of concrete cable-stayed bridges under seismic action. *Computers and Structures*, **222**, pp. 36–47, 2019.

[28] Ferreira, F.L.S. & Simões, L.M.C., Optimum design of active and passive cable stayed footbridges. *Proceedings of the 10th International Conference on Computational Structures Technology*, eds B.H.V. Topping, J.M. Adam, F.J. Pallarés, R. Bru & M.L. Romero, Civil-Comp Press: Stirlingshire, UK, Paper 167, 2010.

[29] Ferreira, F. & Simões, L., Optimum cost design of controlled cable stayed footbridges. *Computers and Structures*, **106–107**, pp. 135–143, 2012.

[30] Ferreira, F. & Simões, L., Optimum design of a controlled cable-stayed footbridge subject to a running event using semiactive and passive mass dampers. *Journal of Performance of Constructed Facilities*, **33**(3), 2019.

[31] Ferreira, F. & Simões, L., Optimum design of a cable-stayed steel footbridge with three dimensional modelling and control devices. *Engineering Structures*, **180**, pp. 510–523, 2019.

[32] Simões, L.M.C. & Negrão, J.H., Optimum design of cable-stayed bridges with imprecise data. *Proceedings of the 8th International Conference on Civil and Structural Engineering Computing*, ed. B.H.V. Topping, Civil-Comp Press: Stirlingshire, UK, Paper 100, 2001.

[33] Negrão, J.H.O. & Simões, L.M.C., Reliability-based optimum design of cable-stayed bridges. *Structural and Multidisciplinary Optimization*, **28**(2–3), pp. 214–220, 2004.

[34] Simões, L.M.C. & Negrão, J.H.O., Reliability-based optimum design of glulam cable-stayed footbridges. *Journal of Bridge Engineering*, **10**(1), pp. 39–44, 2005.

[35] Martins, A.M.B., Simões, L.M.C. & Negrão, J.H.J.O., Optimization of extradosed concrete bridges subjected to seismic action. *Computers and Structures*, **245**, 2021.

[36] Martins, A.M.B., Simões, L.M.C. & Negrão, J.H.J.O., Optimization of under-deck concrete cable-stayed bridges. *Proceedings of the 11th International Conference on Engineering Computational Technology*, eds B.H.V. Topping & P. Iványi, Civil-Comp Press: Stirlingshire, UK, to be published.

[37] Ferreira, F. & Simões, L., Least cost design of curved cable-stayed footbridges with control devices. *Structures*, **19**, pp. 68–83, 2019.

[38] Ferreira, F. & Simões, L., Automated synthesis of controlled cable-stayed footbridges with S-shaped deck. *Advances in Engineering Software*, **149**, 2020.

[39] Ferreira, F. & Simões, L., Optimum seismic design of curved cable-stayed bridges. *Structures*, **43**, pp. 131–148, 2022.

[40] Cid, C., Baldomir, A. & Hernandez, S., Optimum crossing cable system in multi-span cable-stayed bridges. *Engineering Structures*, **160**, pp. 342–355, 2018.

[41] Atmaca, B., Size and post-tensioning cable force optimization of cable-stayed footbridge. *Structures*, **33**, pp. 2036–2049, 2021.

[42] Feng, Y., Lan, C., Briseghella, B., Fenu, L. & Zordan, T., Cable optimization of a cable-stayed bridge based on genetic algorithms and the influence matrix method. *Engineering Optimization*, **54**(1), pp. 20–39, 2022.

[43] Franchini, A., Sebastian, W. & D'Ayala, D., Surrogate-based fragility analysis and probabilistic optimisation of cable-stayed bridges subject to seismic loads. *Engineering Structures*, **256**, 2022.

[44] Cid Montoya, M., Hernández, S. & Nieto, F., Shape optimization of streamlined decks of cable-stayed bridges considering aeroelastic and structural constraints. *Journal of Wind Engineering and Industrial Aerodynamics*, **177**, pp. 429–455, 2018.

[45] Cid Montoya, M., Nieto, F., Hernández, S., Fontán, A., Jurado, J.A. & Kareem, A., Optimization of bridges with short gap streamlined twin-box decks considering structural, flutter and buffeting performance. *Journal of Wind Engineering and Industrial Aerodynamics*, **208**, 104316, 2021.

[46] Cid Montoya, M., Hernández, S. & Kareem, A., Aero-structural optimization-based tailoring of bridge deck geometry. *Engineering Structures*, 114067, 2022.

STRUCTURAL TOPOLOGY OPTIMIZATION WITH HIGH SPATIAL DEFINITION BY USING THE OVERWEIGHT APPROACH

DIEGO VILLALBA, JOSÉ PARÍS, IVÁN COUCEIRO & FERMÍN NAVARRINA
GMNI – Group of Numerical Methods in Engineering, E. T. S. de Ingeniería de Caminos,
Canales y Puertos, Universidade da Coruña, Spain

ABSTRACT

The first formulation of topology optimization was proposed in the 1980s. Since then, many contributions have been presented with the purpose of improving its efficiency and expanding its field of application. The aim of this research is to develop a structural topology optimization algorithm considering minimum weight and stress constraints. Structural topology optimization with stress constraints has been previously formulated with several different approaches, mainly: local stress constraints, global stress constraints or block aggregation of stress constraints. In this research the overweight approach, an improvement of the so-called damage approach, is used. In this method, a virtual relative density (VRD) is defined as a function of the violation of the local stress constraints. VRD is increased as the stresses exceed the maximum allowable value, with the exception of the areas with the minimum value of the relative density, since full-void solutions are intended. The distribution of the material in the domain is modelled using two different approaches: a uniform relative density within each element of the mesh and a relative density defined by means of quadratic B-splines. For this reason, the structural analysis is performed by means of the finite element method (FEM) and the isogeometric analysis (IGA) respectively. The optimization is addressed by means of the sequential linear programming algorithm (SLP), which is driven by the information provided by a full first order sensitivity analysis extension of both FEM and IGA formulations. Finally, the overweight approach is tested by means of some two dimensional problems. The domain has been divided in an elevated number of elements to attain high spatial definition solutions. The results show that the overweight approach is a feasible alternative for the damage approach and the stress constraints aggregation techniques to solve the topology optimization problem. A comparison between both formulations of the material distribution is included.
Keywords: topology optimization, structures, stress constraint, overweight approach, finite element method, isogeometric analysis, aggregation techniques, high spatial definition.

1 INTRODUCTION

One of the first works about structural topology optimization of continuum structures was submitted by Bendsøe and Kikuchi in 1988 [1]. These works supposed the establishment of the basis of this new field. Since then an important number of contributions has been made. Although an important number of different kind of problems have been formulated, only two of them have been deeply analysed: maximum stiffness problem and minimum weight with stress constraints problem. On the other hand, the variety of ways developed to solve these problems is extremely high due to the drawbacks emerged during the solution process. Nevertheless, the attention in this publication will be only focused in the second one because of its high interest since the engineering point of view. The different alternatives used to solve the minimum weight with stress constraints problem are related with the different ways to define the material layout in the domain and the different ways to impose the stress constraints in the topology optimization problem. While the first aspect has a high influence in the quality of the solution regarding to the spatial definition, the second one impacts in the CPU time used to solve the problem. First, regarding to the way to define the material layout in the domain, the approaches developed until now are: the homogenization techniques [2]–

WIT Transactions on The Built Environment, Vol 209, © 2022 WIT Press
www.witpress.com, ISSN 1743-3509 (on-line)
doi:10.2495/HPSU220071

[4], the solid isotropic material with penalty (SIMP) [5], [6], the multimicrostructural approach [7]–[9], the level set methods [10]–[12], the bubble method [13], the phase field approach [14], and more recently the isogeometric analysis [15]–[17]. On the other hand, with regards to the imposition of the stress constraints, the strategies used to impose them until now are: local stress constraints [18], [19], global stress constraints [20], [21], block aggregation of stress constraints [22], [23] and more recently the damage approach [24]. In this research for defining the material layout in the domain the isogeometric analysis will be used. However, the overweight approach, an alternative formulation of the damage approach previously developed, is established to impose the stress constraints in the topology optimization problem.

2 STRUCTURAL ANALYSIS

The structural analysis used in the topology optimization problem proposed is the classical method of the finite element method. In this approach several hypotheses must be considered: small displacements, small displacement derivatives and elastic and linear material. On the other hand, this formulation is also used with the isogeometric analysis. Since the design variables used in the topology optimization problem define the material layout, it is necessary to consider the effect of the relative density in the finite element formulation. For this reason, the classical finite element method formulation has to be modified in order to include the design variables of the optimization problem. This issue has been addressed by using a numerical formulation that includes the effect of the relative density [22], [25], [26]. This formulation is equivalent in a certain sense to the SIMP method, and only requires to introduce slight modifications in the classical formulation. These modifications are reduced to consider the effect of the relative density in the integrals of the contribution of each finite element or knot span of the mesh, in case of using the finite element method or the isogeometric analysis respectively. Therefore, the structural analysis can be obtained by solving the system of linear equations:

$$K\alpha = f, \tag{1}$$

where in this case:

$$K \equiv K(\rho), \quad \alpha \equiv \alpha(\rho), \quad f \equiv f(\rho). \tag{2}$$

The matrix K is the structural stiffness matrix, α is the nodal displacement vector and f is the applied loads vector. Each one of the terms K_{ji} of the stiffness matrix K is obtained as:

$$K_{ji} = \sum_{e=1}^{N_e} K_{ji}^e, \quad j = 1, \dots, N, \quad i = 1, \dots, N, \tag{3}$$

where N_e is the number of elements or knot spans, depending on the method used to compute the structural analysis, finite element method or isogeometric analysis, respectively; and N is the number of the nodes of the mesh. At this point, it is important to remark that the number of points required to compute the structural analysis is different depending on the method used, since the way to define the material layout will be also different. Except for particular cases, isogeometric analysis needs fewer nodes than the finite element method. The elemental stiffness matrixes K^e must be computed as usual by multiplying the corresponding integrand times the relative density of element or knot span e [25], [26]. On the other hand, the applied forces vector f_j can be obtained as usual in a conventional finite element formulation by considering that the contribution of the forces per unite of volume must be multiplied times the relative density [25], [26]. The whole finite element formulation including the relative density effect has been specifically addressed in previous papers by Navarrina et al. [25] and

Paris et al. [26]. The linear equation system proposed in eqn (1) is solved by means of a Cholesky factorization algorithm since it allows to compute additional systems of equations with small effort. These additional linear equation systems are required in the attainment of the sensitivity analysis. Finally, if a comparative analysis between the use of the finite element method or the isogeometric analysis is made, it is expectable that the CPU time requirements will be lower in the case of the isogeometric analysis, since the size of the equation system is lower due to the number of points required to compute it.

3 OPTIMIZATION PROBLEM

The optimization problem can be formulated in different ways; however, its more typical formulation is:

$$
\begin{aligned}
&Calculate && \rho = \{\rho_i\}, && i = 1, \dots, n \\
&which \left(\frac{minimize}{maximize}\right) && F(\rho) \\
&verifying && g_j(r_j^0, \rho) \le 0 && j = 1, \dots, m && (4) \\
& && h_l(r_l^0, \rho) = 0 && l = 1, \dots, p \\
& && \rho_{min} \le \rho_i \le \rho_{max} && i = 1, \dots n,
\end{aligned}
$$

where $\rho=\{\rho_i\}$ is the design variable vector, $F(\rho)$ is the objective function, g_j are the inequality constraints of the problem and, finally, h_l are the equality constraints of the problem. Furthermore, in the optimization problem, n is the number of design variables, while m and p are, respectively, the number of inequality and equality constraints. Finally, as far as optimization problem concerns, it will be also necessary to establish the side constraints of the design variables, which are respectively ρ_{min} and ρ_{max}.

3.1 Objective function

In the topology optimization problem developed in this publication, the objective function of the problem is the structural weight whose value will have to be minimized. The formulation of the objective function will be, therefore:

$$F(\rho) = \sum_{e=1}^{N_e} \int_{\Omega_e} \rho \, d\Omega_e, \qquad (5)$$

where N_e is the number of elements or knot spans in which the domain is discretised, Ω_e represents the domain of the element or knot span e, and ρ is the relative density. At this point, the methods used to define the material layout are briefly explained. First, each design variable will represent the relative density in each element of the mesh in the finite element formulation, however, in the isogeometric analysis formulation, the scheme used to define the density relative coincides with the scheme used to compute the structural analysis, for this reason, it is necessary to use the same B-splines, since the design variables are in this case, the value of the relative density in the control points. Finally, a factor which penalizes intermediate relative density values "p" is introduced in the formulation since the attainment of full-void solutions is intended. This factor "p" is introduced in the objective function, and as a result, the objective function of the topology optimization problem will be:

$$F(\rho) = \sum_{e=1}^{N_e} \int_{\Omega_e} \rho^{1/p} d\Omega_e. \qquad (6)$$

3.2 Overweight constraint

In the topology optimization problem developed in this publication, there will be only one inequality constraint. This constraint will be known as overweight constraint, since it is obtained through the application of the overweight approach. The overweight approach is used as an alternative way to impose the stress constraints in the topology optimization problem. In this approach, two different models that describe the same mechanical body will have to be defined. They will be known hereinafter as original model and overweight model. The overweight model will be defined from the original one doing a sequence of steps. First, the structural analysis of the original model is computed. Then, the structural stresses in the points where they have to be checked can be computed. Finally, an overweight is introduced in the model in the areas where stresses exceed their maximum allowable value. Since the relative density is the property modified in the original model to create the overweight model, the relative density in the overweight model will have to satisfy these conditions:

$$\begin{cases} \tilde{\rho}(x) > \rho(x) & \forall x \in \Omega_\sigma := \{x \mid |\sigma(x)| > \sigma_{lim}\} \\ \tilde{\rho}(x) = \rho(x) & \forall x \in \Omega \backslash \Omega_\sigma, \end{cases} \tag{7}$$

where ρ is the relative density in the original model and $\tilde{\rho}$ in the overweight model. Finally, the overweight constraint imposed in the topology optimization problem is a comparison between the structural weight of both models:

$$g(\rho) = \frac{\tilde{W}}{W} - 1 \leq 0, \tag{8}$$

where W is the structural weight of the original model and $\tilde{W}$ is the structural weight of the overweight model and they can be computed as follows:

$$W(\rho) = \int_\Omega \rho \, d\Omega \qquad \tilde{W}(\rho) = \int_\Omega \tilde{\rho} \, d\Omega. \tag{9}$$

3.2.1 Overweight model
The formulation of the relative density in the overweight model can be defined by different ways. A similar scheme that in Verbart et al. [24] is used to define the relative density in the overweight model, this relative density is increased in the overweight model. Furthermore, the relative density of the overweight model can be computed as:

$$\tilde{\rho} = \rho_{min} + \beta(\rho - \rho_{min}) \qquad where \ \beta(\sigma; \sigma_{lim}) \geq 1, \tag{10}$$

where β is the overweight function. This function has to satisfy two conditions: to be at least first order differentiable and to be monotonically crescent when stresses exceed its allowable limit.

3.3 Side constraints

The side constraints of the design variables are the only term of the topology optimization problem which has not been defined yet. While the upper limit of the design variables is obvious since its value will be equal to 1, what represents the area full of material, the lower limit of the design variables will not be null, since this circumstance can produce inherent problems in some parts of the topology optimization algorithm, like the singularity of some matrixes. For this reason, the value of this lower limit will have to be a value slightly higher to zero; in this publication it will be equal to 0.001.

4 OPTIMIZATION ALGORITHMS

The solution of the topology optimization problem (4) will be made by means of the use of different algorithms. The algorithms used to solve the problem in this publication are: the steepest descent method, the sequential linear programming algorithm based on the simplex algorithm and a, so called, back to the feasible region algorithm based on constraint derivatives. In the first place, the steepest descent method will be only used when the value of all the structural stresses considered are inferior to a certain percentage of the maximum allowable value. This circumstance only happens in the first iteration of the topology optimization problem, since the structural domain is full of material in the first iteration. For this algorithm, it is only necessary to compute the objective function sensitivity analysis. On the contrary, the back to the feasible region algorithm is used when the structural weight of the overweight model exceeds a certain percentage of the structural weight of the original one. In this case, the solution is far from the feasible region, and the simplex algorithm does not run properly. For this reason, it is only necessary to compute in this case the overweight constraint sensitivity analysis. Finally, in the intermediate situations the simplex algorithm is used. This situation will be more common during the optimization procedure since the solution tend to be near to the border of the feasible region. Therefore, in this case it is necessary to compute both sensitivity analyses: objective function and overweight constraint. The main reason to use these algorithms is to compute the best improvement direction at each iteration depending on the situation of the previous solution. Therefore, once the improvement direction has been computed, the next step is to establish the most appropriated value of the improvement factor which will determine the solution of this iteration. For this purpose, the first order Taylor series will have to be calculated. On the other hand, a linear search will be made in order to avoid the calculus of directional derivatives, since the first order directional derivatives can be calculated directly with the information used with the previous algorithms. Lastly, once the improvement direction and the improvement factor are known, the next step is to update the solution of the topology optimization problem, and to repeat the optimization process until convergence.

5 SENSITIVITY ANALYSIS

The solution of the topology optimization problem requires to compute the derivatives of the objective function and the overweight constraint. The process of computing the derivatives is known as sensitivity analysis. This procedure is the other key point in the solution of the topology optimization problem apart from the structural analysis. In this publication only a first order sensitivity analysis will be developed. On the other hand, it is necessary to use two different approaches for doing the first order sensitivity analysis due to the different characteristics of the objective function and the overweight constraint. First, the sensitivity analysis of the objective function can be computed analytically, since the structural weight depends directly on the design variables. On the contrary, the calculation of the sensitivity analysis of the overweight constraint require the use of more complex procedures, since the relationship between the damage constraint and the design variables is not as straightforward as in the objective function. On the other hand, there are other approaches more computationally efficient than the direct differentiation, if the number of constraints is considerably lower than the number of design variables. This circumstance happens in this problem where only one constraint is defined. For this reason, the approach used in this publication to calculate the sensitivity analysis of the overweight constraint will be the adjoint variable approach [27]. Although, the directional derivatives of the objective function and the overweight constraints are necessary for the optimization algorithm, their calculation can be made directly from the first order derivatives of them, by means of the multiplication of the first order derivative vector times the improvement direction obtained with the algorithms previously introduced. Finally, it is important to remark an important difference between the use of the finite element method and the isogeometric analysis regarding to the way that each design variable have influence in one or more elements or knot spans. While for the finite element method used in this publication, only one design variables has influence in each element of the mesh, for the isogeometric analysis proposed in this publication, this number will be equal to NINE. These numbers depend on the interpolation scheme used with the isogeometric analysis. Consequently, if the interpolation scheme is changed, the number of design variables which has influence in each element or knot span can be not only increased but also reduced. Because of this circumstance, an increase of the computational requirements and CPU time will be obtained in case of using the isogeometric analysis instead of the finite element method. Furthermore, this situation supposes a contradiction with respect to the structural analysis, where the optimal method to do it was the isogeometric analysis.

6 APPLICATION EXAMPLES

In this paper, only one example is shown. This example is used to test the performance of the overweight approach technique as an indirect way to impose the stress constraints. This example is solved both with the finite element method and the isogeometric analysis. The numerical example solved in this paper is the optimum design of a cantilever beam. The proposed example corresponds to a 2D structure in plane stress. Finally, a comparative analysis of the results with both methods is made, not only regarding to the quality of the solutions but also the computational requirements used to obtain them. This analysis let establish if one of the methods is highly advisable to solve the topology optimization problem or if there are not important differences between them.

6.1 Optimum design of a cantilever beam

The example studies the optimum design of a cantilever beam with null displacements in the left edge and with a vertical force applied in the middle of the right edge. Fig. 1 shows the dimensions of the domain and the position of the vertical force applied. Self-weight of the structure is included as a structural load in this example. The domain of the structure has been discretized by using a homogeneous mesh with $160 \times 80 = 12,800$ 8-node quadrilateral elements in case of the finite element method and quadratic knot spans in case of the isogeometric analysis. The thickness of the structure is 0.2 m. The external force applied (1×10^3 kN) has been distributed in eight contiguous elements or knot spans in order to avoid stress accumulation phenomena. The material being used in this problem is steel with density $\gamma_{mat} = 7,850$ kg/m^3, Young's Modulus $E = 2.1 \times 10^5$ MPa, Poisson ratio $v = 0.3$ and elastic limit $\sigma_{max} = 230$ MPa. Fig. 2 shows the optimal solution of the problem for the finite element method and the isogeometric analysis. The optimal solution is represented by means of its material distribution in the domain with the value of the relative density at each point. This relative density defines the percentage of the material phase. The structural weight of the optimal solution is 136.75 kg in case of the finite element method and 130.02 kg in case of the isogeometric analysis, this means the 17.42% and the 16.56% of the structural weight of the initial solution (all the domain is full of material, structural weight of the initial solution: 785 kg).

Fig. 3 shows the normalized stress state for the cantilever beam problem obtained by using the overweight approach for both methods. This normalized stress σ_e* has been computed by

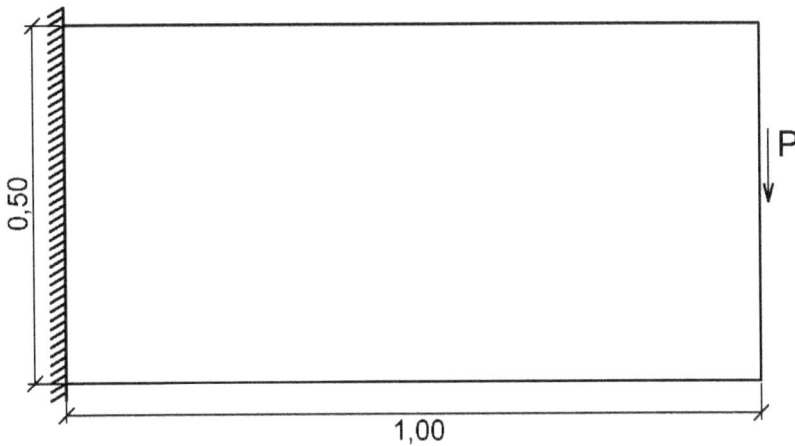

Figure 1: Scheme of the cantilever beam problem (units in m).

means of the maximum allowable value of the stress σ_{max}, the relaxation parameter of stresses φ_e which increases the value of the maximum allowable stress when the relative density is lower than 1 and the Von Mises stress $\sigma_{VM,e}$, which is the most typical failure criterion when the steel is the structural material.

The optimal solution of the topology optimization problem has been obtained after 2250 iterations with both methods: finite element method and isogeometric analysis. Finally, the distribution of the computing time per algorithm of all the iterations computed in the optimization process for both methods: finite element method and isogeometric analysis is shown in Tables 1 and 2, respectively.

6.2 Comparative analysis

In the first place, the way to define the relative density in the domain plays an important role in the obtained results. In general, the results obtained with the isogeometric analysis proposed are better in terms of quality in comparison with the obtained with the finite element method proposed, regarding to the spatial definition. In other words, the number of design variables used to obtain solutions with the same spatial definition is lower with the isogeometric analysis proposed. Nevertheless, this circumstance can be solved, if an alternative way to define the material layout is established in case of the finite element method. However, the topology of the results obtained with both methods is the same, since the optimal solution consists on a set of bars that coincide with the isostatic lines. In the second place, regarding to the CPU time, the critical algorithm of both methods does not coincide. While the structural analysis is the critical step of the finite element method proposed, since it supposes the 75% of the time required to solve the topology optimization problem, the sensitivity analysis is the critical part of the isogeometric analysis, since it supposes the 60% of the time used to solve the topology optimization problem. On the one hand, the CPU time required to compute the structural analysis is related with the number of points used to solve it, since this number is related with the size of the structural problem. On the other hand, the CPU time required to compute the sensitivity analysis is related with the number of design variables which have influence in each element of the mesh or knot span

(a)

(b)

Relative Density

1.0e-03 0.1   0.2   0.3   0.4   0.5   0.6   0.7   0.8   0.9 1.0e+00

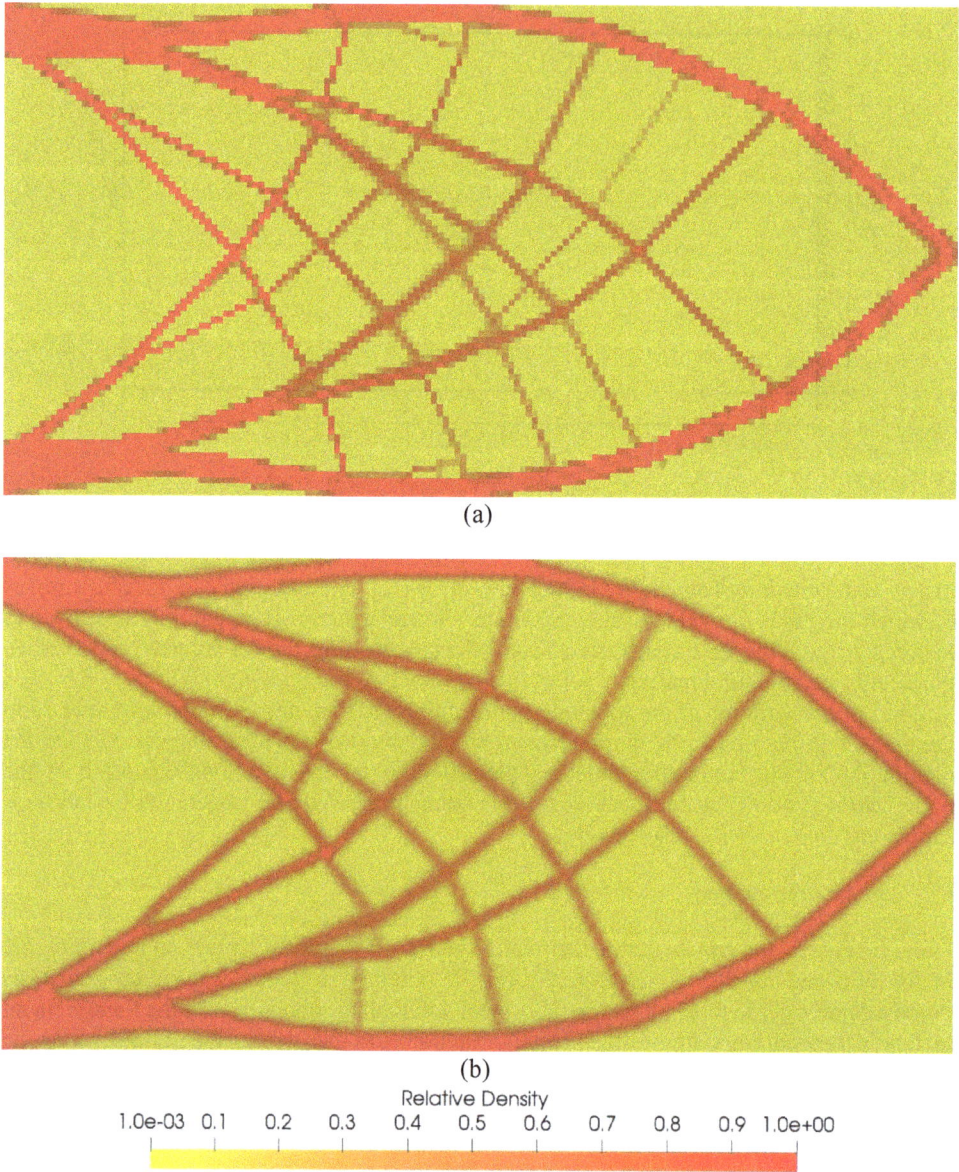

Figure 2:   Optimal solution of the cantilever beam problem by using overweight approach.
            (a) Finite element method; and (b) Isogeometric analysis.

of the patch, since this number is related with the way to define the material layout in the domain. Finally, other important aspect that does not suppose an important difference between both methods, is the appearance of stresses whose value is higher than their maximum allowable value. This circumstance happens in the areas where there is a stress concentration phenomenon, and in areas whose relative density is close to their minimum allowable value.

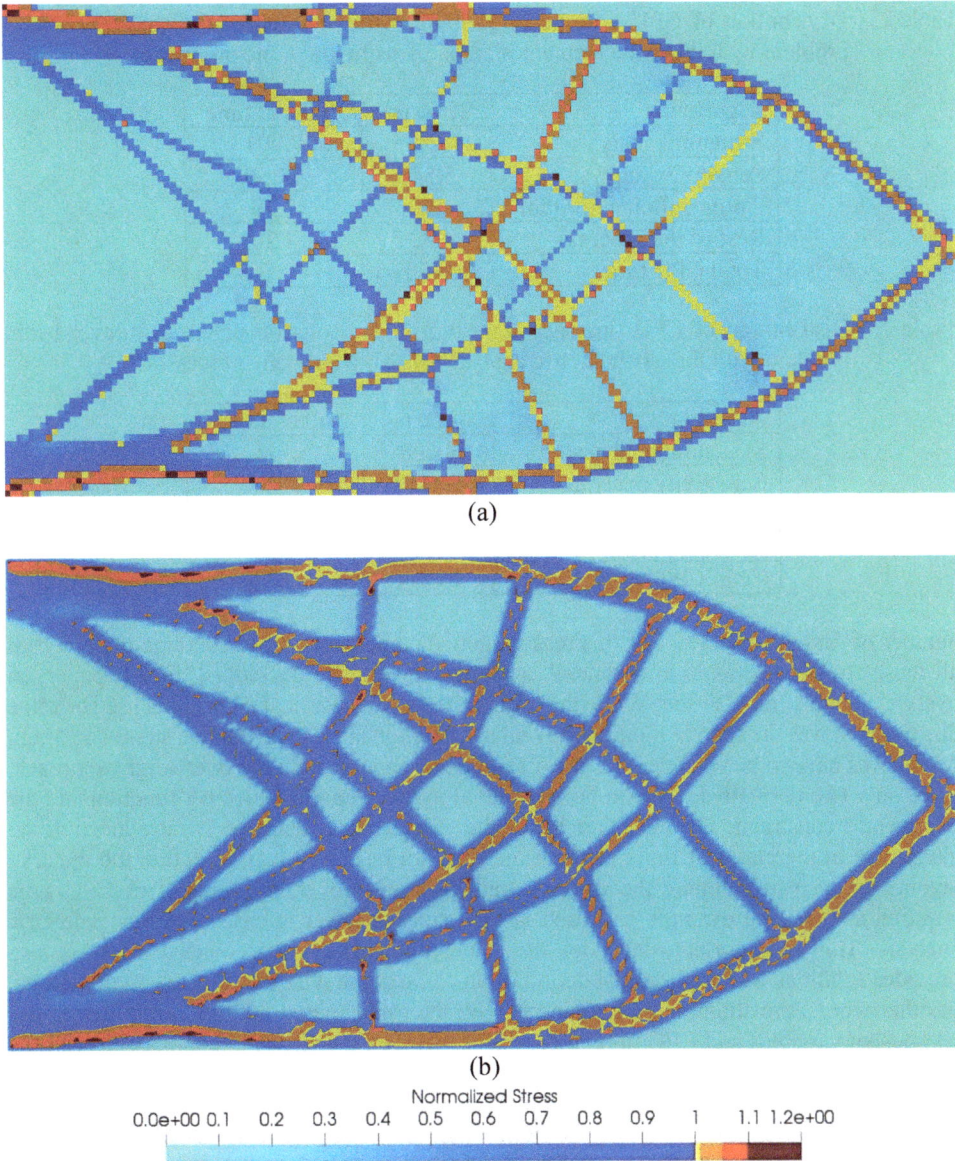

Figure 3:   Normalized stress state of the cantilever beam problem by using the overweight approach ($\sigma_e^* = \sigma_{VM,e}/(\varphi_e \, \sigma_{max,})$). (a) Finite element method; and (b) Isogeometric analysis.

7 CONCLUSIONS

This paper introduces a new method to solve the topology optimization of structures problem with minimum weight and stress constraints. The attainment of results with high spatial definition makes necessary to consider a large number of stress constraints and to use a large

Table 1: Distribution of CPU time per algorithm in the solution of the cantilever beam problem with the finite element method by using only one processor.

Algorithms	Time (s)	% Total time
Structural analysis	17,235	73.01
Sensitivity analysis	5,546	23.50
Optimization algorithm	539	2.29
Rest of the process	283	1.20
Total CPU time	23,603	

Table 2: Distribution of CPU time per algorithm in the solution of the cantilever beam problem with the isogeometric analysis by using only one processor.

Algorithms	Time (s)	% Total time
Structural analysis	4,608	27.41
Sensitivity analysis	10,418	61.97
Optimization algorithm	702	4.18
Rest of the process	1,083	6.44
Total CPU time	16,811	

number of design variables. For this reason, new methods which can combine the effect of all these stress constraints in a general constraint have to be developed. In this paper, an overweight constraint is used. While the critical step of the classical algorithms used to solve this problem was the search of the improvement direction since all the active stress constraint derivatives have to be considered. This drawback disappears with the overweight approach, since only two derivatives have to be considered in this step: the objective function and the overweight constraint. On the other hand, the isogeometric analysis is introduced as an alternative to the classical finite element method, not only as a way to define the material layout, but also to compute the structural analysis. Moreover, the amount of CPU time required to solve the example proposed in this paper is lower in case of the isogeometric analysis. This is important since, for the same problem, the isogeometric analysis also provides solutions with high spatial definition in comparison with the finite element method. In other words, the attainment of solutions with the same spatial definition requires to use less design variables and CPU time in case of isogeometric analysis in comparison with the finite element method. In conclusion, the use of the Overweight Approach reports important benefits from a computational point of view in topology optimization problems with stress constraints, and the results obtained are similar in terms of quality with the obtained with the classical approaches. The efficiency of the isogeometric analysis in comparison to the finite element method has been also clearly demonstrated, since solutions with high spatial definition are obtained in less CPU time.

ACKNOWLEDGEMENTS
This work has received financial support from the Xunta de Galicia (Secretaria Xeral de Universidades) and the European Union (European Social Fund (ESF)) through "Grants for funding predoctoral stages in universities, public research organisms and other R&D entities of Galicia" ED481A-2016/387. This work has also been partially supported also by FEDER funds of the European Union, by the "Ministerio de Economía y Competitividad" of the Spanish Government through grants DPI2015-68431-R and RTI2018-093366-B-I00 and by the "Consellería de Educación, Universidade e Formación Profesional" of the Xunta de

Galicia, through grants for the consolidation and structuring of competitive research units of the Galician University System: GRC2014/039 and GRC2018/41.

REFERENCES

[1]   Bendsøe, M.P. & Kikuchi, N., Generating optimal topologies in structural design using a homogenization method. *Computer Methods in Applied Mechanics and Engineering*, **71**, pp. 197–224, 1988.

[2]   Murat, F. & Tartar, L., Calcul des variations e homogénéisation, les méthodes de l'homogénéisation théorie et applications en physique. *Collegue Dir. Etudes et Recherches EDF*, **57**, pp. 319–369, 1985.

[3]   Suzuki, K. & Kikuchi, N., A homogeneization method for shape and topology optimization. *Computer Methods in Applied Mechanics and Engineering*, **93**(3), pp. 291–318, 1991.

[4]   Allaire, G., Jouve, F. & Maillot, H., Topology optimization for minimum stress design with the homogenization method. *Structural and Multidisciplinary Optimization*, **28**, pp. 87–98, 2004.

[5]   Rossow, H.P. & Taylor, J.E., A finite element method for the optimal design of variable thickness sheets. *AIAA Journal*, **11**, pp. 1566–1569, 1973.

[6]   Bendsøe, M.P. (ed.), *Optimization of Structural Topology, Shape and Material*, Springer-Verlag, 1995.

[7]   Rodrigues, H., Guedes, J.M. & Bendsøe, M.P., Hierarchical optimization of material and structure. *Structural Multidisciplinary Optimization*, **24**, pp. 1–10, 2002.

[8]   Nakshatrala, P.B., Tortorelli, D.A. & Nakshatrala, K., Nonlinear structural design using multiscale topology optimization. Part I: Static formulation. *Computer Methods in Applied Mechanics and Engineering*, **261–262**, pp. 167–176, 2013.

[9]   Sivapuram, R., Dunning, P.D. & Kim, H.A., Simultaneous material and structural optimization by multiscale topology optimization. *Structural Multidisciplinary Optimization*, **54**, pp. 1267–1281, 2016.

[10] Sethian, J.A. & Wiegmann, A., Structural boundary design via level set and immersed interface methods. *Journal of Computational Physics*, **163**, pp. 489–528, 2000.

[11] Wang, M.Y., Wang, X. & Guo, D., A level set method for structural topology optimization. *Computer Methods in Applied Mechanics and Engineering*, **192**, pp. 227–246, 2003.

[12] Amstutz, S. & Andrä, H., A new algorithm for topology optimization using a level-set method. *Journal of Computational Physics*, **216**, pp. 573–588, 2006.

[13] Eschenauer, H.A. & Schumacher, A., Bubble method: A special strategy for finding best possible initial designs. *Proceedings of the ASME Design Technical Conference 19th Design Automotion Conference*, **65**, pp. 437–443, 1993.

[14] Wang, M.Y. & Zhou, S., Phase field: A variational method for structural topology optimization. *Tech Science*, **6**(6), pp. 547–566, 2004.

[15] Hassani, B., Khanzadi, M. & Tavakkoli, S.M., An isogeometrical approach to structural topology optimization by optimality criteria. *Structural Multidisciplinary Optimization*, **45**, pp. 223–233, 2012.

[16] Qian, X., Topology optimization in B-spline space. *Computer Methods in Applied Mechanics and Engineering*, **265**, pp. 15–35, 2013.

[17] Lieu, Q.X. & Lee, J., A multi-resolution approach for multi-material topology optimization based on isogeometric analysis. *Computer Methods in Applied Mechanics and Engineering*, **323**, pp. 272–302, 2017.

[18] Duysinx, P. & Bendsøe, M.P., Topology optimization of continuum structures with local stress constraints. *International Journal of Numerical Methods in Engineering*, **43**, pp. 1453–1478, 1998.

[19] Navarrina, F., Muiños, I., Colominas, I. & Casteleiro, M., Minimum weight with stress constraints topology optimization. *Proceedings of 7th US National Congress on Computational Mechanics*, 2003.

[20] Martins, J.R.R.A. & Poon, N.M.K., On structural optimization using constraint aggregation. *VI World Congress on Structural and Multidisciplinary Optimization*, 2005.

[21] Poon, N.M.K. & Martins, J.R.R.A., An adaptive approach to constraint aggregation using adjoint sensitivity analysis. *Structural Multidisciplinary Optimization*, **34**, pp. 61–73, 2007.

[22] Paris, J., Navarrina, F., Colominas, I. & Casteleiro, M., Block aggregation of stress constraints in topology optimization of structures. *10th International Conference on Computer Aided Optimum Design in Engineering, OPTI 2007*, 2007.

[23] Paris, J., Navarrina, F., Colominas, I. & Casteleiro, M., Advances in the statement of stress constraints in structural topology optimization. *4th International Conference on Advanced Computational Methods in Engineering, ACOMEN 2008*, 2008.

[24] Verbart, A., Langelaar, M. & Van Keulen, F., Damage approach: A new method for topology optimization with local stress constraints. *Structural Multidisciplinary Optimization*, **53**, pp. 1081–1098, 2016.

[25] Navarrina, F., Muiños, I., Colominas, I. & Casteleiro, M., Topology optimization of structures: A minimum weight approach with stress constraints. *Advanced Engineering Software*, **36**, pp. 599–606, 2005.

[26] Paris, J., Navarrina, F., Colominas, I. & Casteleiro, M., Topology optimization of continuum structures with local and global stress constraints. *Structural Multidisciplinary Optimization*, **39**(4), pp. 419–437, 2009.

[27] Paris, J., Restricciones en tension y minimización del peso: Una metodología general para la optimización topológica de estructuras, PhD thesis, University of A Coruña, 2007.

SECTION 3
BLAST AND IMPACT LOADS

ANALYSIS OF CHARGE SHAPE INFLUENCE ON BLAST PRESSURE

HRVOJE DRAGANIĆ, SANJA LUKIĆ, IVAN RADIĆ, GORAN GAZIĆ & MARIO JELEČ
Faculty of Civil Engineering and Architecture Osijek, Josip Juraj Strossmayer University of Osijek, Croatia

ABSTRACT

Blast wave intensity depends on several parameters, namely: explosive material type, charge weight, shape and orientation, detonation point position, detonation initiator type (primary explosive type), the position (distance) of the explosive charge in relation to the intended target (standoff distance) and ground surface. Environmental conditions, particularly air temperature, humidity and atmospheric pressure, also influence blast pressures. It is difficult to accurately predict the blast wave action on target structures if all of these parameters are considered. This research concentrates on the influence of the shape of the explosive charge on blast pressure measurements. Spherical and hemispherical charge shapes are considered usual and, as such, accurate and reliable analytical expressions for the blast wave pressure approximation are available. The form and propagation of spherical charge blast waves are considered to have been thoroughly studied and known. In today's urban and guerrilla warfare, speed of action is a crucial factor. Rendering the careful shaping of explosive charges is time consuming and unnecessary, hence the need for investigating different charge shapes, other than spherical. This investigation consisted of field range experimental measurements of the incident (free-field) and reflected pressures caused by detonating differently shaped charges. The shapes considered were: spherical, cylindrical and rectangular. The experiments were numerically validated and verified using ANSYS Autodyn hydrocode software. Numerical simulations utilised a coupled Euler–Lagrange planar solver, using an ideal air environment and PEP500 explosive material. Charge shapes varied, according to the experimental outline, and the measuring points were constant, to allow comparison of the measured data.
Keywords: blast load, charge shape, incident pressure, ANSYS Autodyn.

1 INTRODUCTION

Analytical expressions provide a satisfactory pressure approximation of spherical and hemispherical charges. The propagation of blast waves induced by the detonation of spherical and hemispherical charges is straightforward. The blast wave pressure front expands radially in all directions, as concentric circles from the detonation point. Most research into the blast load on structures is concerned with the blast load produced by spherical charges. Due to thorough research, the blast load from spherical explosive charges is easy to calculate and simulate. Blast load mitigation usually deals with this kind of charge. Today, however, urban and guerrilla warfare mandates speed of action, and so the careful shaping of explosive charges is obsolete and time-consuming. Speed results in arbitrary charge shapes that produce blast pressures and wave propagation, which are different from spherical explosive charges. One of the charge shapes studied in more detail is a cylindrical charge. Recent research emphasises length to base diameter (L/D) as being the main parameter which influences the blast wave shape and pressure distribution. Wu et al. [1] studied the effect of L/D, charge orientation, and scaled distance on blast parameters from the detonation of cylindrical charges, utilising finite element analysis and a small number of experiments. The results indicate that peak reflected pressures and impulses from a vertically-orientated cylinder (with an axis perpendicular to the span of the target) were much larger than those from a spherical charge of equal mass, or a horizontally-orientated cylinder. Sherkar et al. [2], Artero-Guerrero et al. [3] and Langran-Wheeler et al. [4], [5] reported similar findings. Cylindrical blast parameters are dependent on the detonation initiation point. Studies have shown that pressure can be enhanced by up to three times for centrally-detonated cylinders [6], [7]. Nevertheless, there is a current lack of understanding of the early and mid-stages of blast wave propagation following the detonation of non-spherical charges.

This investigation consisted of field range experimental measurements of incident (free-field) and reflected pressures resulting from the detonation of differently shaped charges. Experiments were numerically simulated using the ANSYS Autodyn® hydrocode software.

2 EXPERIMENT

Experiments were conducted on a military field range, to ensure a safe and secure environment. Military personnel were in charge of explosives handling, charge shaping, placement and detonation. The charges were placed vertically in the tests. Two blast wave parameters, the blast wave incident and reflected pressure, were measured and consequently compared to results obtained by numerical modelling and the empirical expressions derived by Kingery and Bulmash [8].

2.1 Description of explosive

Lead azide detonator capsules initiated the explosive reaction of a PEP500 charge in the field range tests [9]. Lead azide is a primary explosive and it is used as a detonator because it has a very stable shelf life. PEP500 is a secondary plastic explosive with a density of 1.5 g/cm^3 and a detonation velocity of 7,400 m/s. Plastic explosives can be formed in arbitrary shapes with no danger of losing their stability and triggering an accidental explosion. The shape of the PEP500 charge varied in three shapes: spherical, cylindrical and rectangular. Each test consisted of detonating 100 g of plastic explosives and pressure measurements were taken at the same standoff distance. Table 1 lists the charge geometries. Fig. 1 shows the explosive packaging, charge shapes and detonator capsules. Expert personnel of the Croatian army conducted the experimental testing at the field range.

Table 1: Charge shape geometry.

Charge shape	Shape dimensions (mm)							
	Height (h)		Width (w)		Length (l)		Base radius (r)	
	M	C	M	C	M	C	M	C
Spherical	/	/	/	/	/	/	27.5	25.0
Cylindrical	100.0	94.3	/	/	/	/	15.0	15.0
Rectangular	30.0	33.0	40.0	40.0	50.0	50.0	/	/

Note: M = field measured dimensions; C = calculated for numerical modelling from mass equation.

2.2 Experimental setup

The field tests consisted of detonating 100 g of PEP500 explosives in different shapes and at a distance of 1.857 m from pressure sensors. Remote detonation utilised an electric discharge from a safe cover, 50 m away. The detonator was placed vertically inside a charge to minimise the influence of positioning on the generated pressures. The cover consisted of the data acquisition system for pressure measurements and the remote detonator. The pressure sensor distance was kept constant throughout each subsequent detonation, to maintain consistency in the measured data. The blast wave incident and reflected pressures were

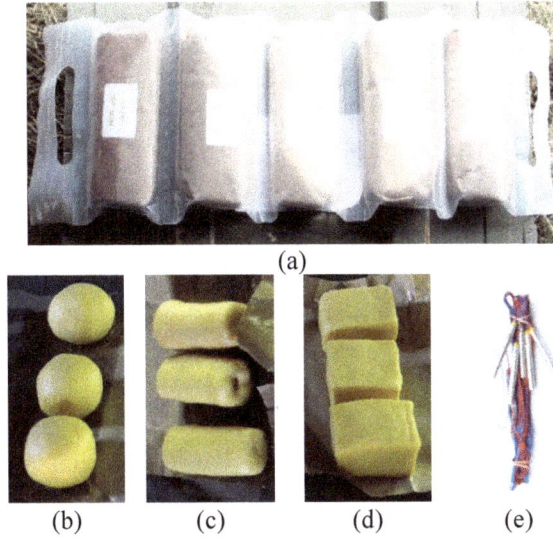

Figure 1:   Charge shapes and detonator capsules. (a) Original packaging of plastic explosive PEP500; (b) Spherical; (c) Cylindrical; (d) Rectangular; and (e) Detonator.

measured. The test implemented four quartz free-field ICP® blast pressure pencil probes (PCB 137B23B), to measure incident pressures, and two IEPE® reflected pressure sensors (Dytran IEPE 2300 V6) for measuring reflected pressure. Pencil probes were placed at an equal distance from the centre of the charge in all four directions. Pencil probes were designated as 11126, 11127, 11128 and 11129. Reflected pressure sensors, designated as 5770 and 5771, were placed at an angle of 225°. A wooden stand held the charge at 1.0 m height in all tests. The sensors were placed at the same elevation and oriented to the charge centre. The positions of the probes are shown in Fig. 2. The field range test setup is shown in Fig. 3. Each charge shape was detonated four times, to obtain a larger data set of measured pressures for comparison and analysis. The weight of the explosive is labelled as W, the actual distance from the centre of the explosive to the sensor is R, and the scaled distance is Z. Considering the distance and charge weight, the adopted scaled distance was 4.0 m/kg$^{1/3}$.

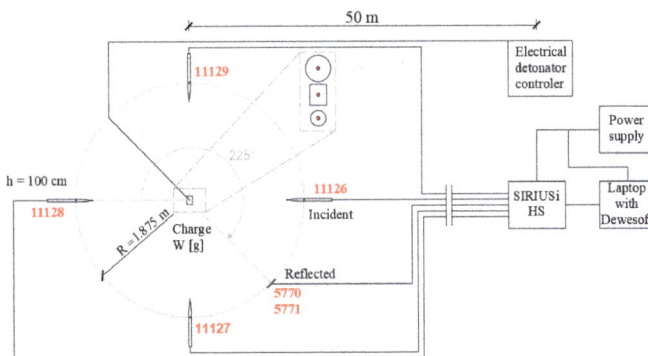

Figure 2: Test setup disposition (location of pressure sensors).

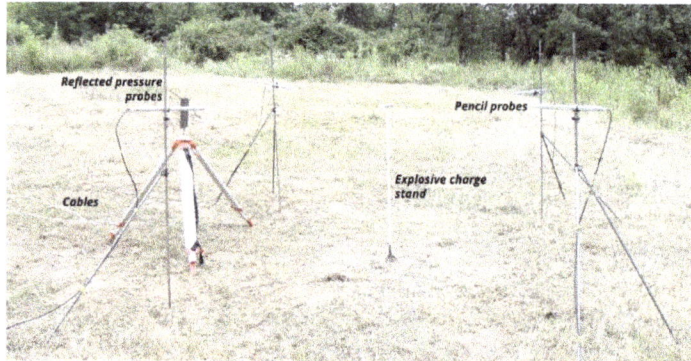

Figure 3: Field range test setup.

Based on previous experience in field blast tests [10], to prevent fireball influence, the charges were not placed too close to the sensors. Fireballs can affect sensors at a distance of 0.835 m ($Z = 1.8$ m/kg$^{1/3}$), so these tests utilised larger distances.

3 NUMERICAL MODELLING

Simulations of experimental tests utilised the ANSYS Autodyn hydrocode software [11]. The numerical models used a planar multi-material Euler Godunov solver [12] and the use of symmetry enabled faster calculation. The model consisted of two parts: air and explosive charge. Air was modelled as an ideal gas while the explosive was PETN1.5, via the Jones–Wilkins–Lee (JWL) equation of state [13]. Air is a gaseous material and solely serves as a transfer medium for the blast waves generated by detonation. The explosive does not have any strength law but it is essential for generating blast waves. Based on the available ANSYS library materials, it was concluded that the PETN1.5 explosive was the one most similar to PEP500.

Table 1 lists the charge shapes and geometrical dimensions used in the field tests. Fig. 1 shows the geometry of the charges. Based on weight, the specific dimensions were calculated and compared to the measured values used in the experiment. The charge weight was considered a constant value throughout all field tests. The calculated geometry parameters for all three charges were approximately the same as the charges detonated, resulting in the minimal influence of charge geometry on the measured blast pressures in the numerical simulations.

A numerical model of the experimental setup is shown in Fig. 4. The symmetry condition allows the modelling of only half of the air environment. The flow out boundary condition is defined on air environment edges to enable free flow of the blast pressures outside the defined air environment, avoiding pressure reflection and amplification. Gauge points 1 (11126), 2 (11127 and 11129) and 4 (11128) measure the incident pressures. As in the experiment, the steel plate is modelled behind gauge point 3 (5770 and 5771), to enable blast pressure reflection and measurement of the reflected pressure. The model setup was similar for all the tests, varying the charge shapes defined in Table 1. A half-circle PETN1.5 fill in an air environment represents a spherical charge, as does the vertically oriented cylindrical charge but with an appropriate base radius. A rectangular PETN1.5 fill with rectangular sides, oriented in the same way as in the field tests, represents a rectangular charge. Length is oriented vertically and height/thickness is oriented laterally. The detonation point is defined in the middle of all simulated charge shapes.

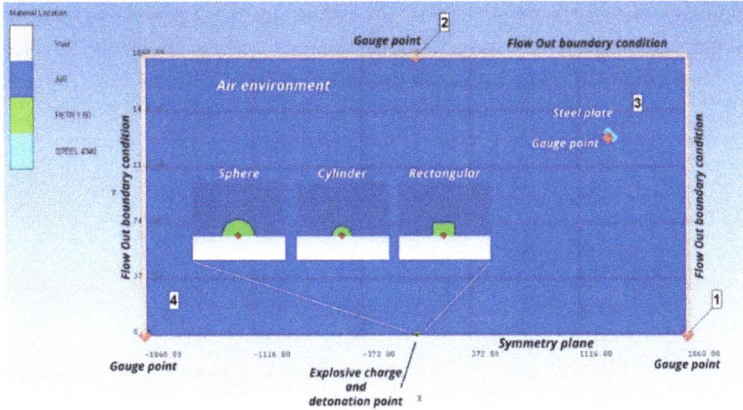

Figure 4: Numerical model of experimental setup.

4 ANALYTICAL CALCULATION

The Kingery and Bulmash (K&B) expressions estimated field pressures for the spherical charge in free-air and as surface burst. The pressure calculation considers both burst types because the reflection of the blast wave from the ground is expected, due to the greater distance of the measuring instruments from the charge (1.857 m) than from the charge to the ground (1.0 m). Incident and reflected pressures were calculated for 100 g of PEP500 explosive (TNT equivalent 1.3). These equations are the basis for predicting pressures and are integrated into various software, such as CONWEP [14]. However, this analysis aims to determine the differences in pressure; do the expressions overestimate or underestimate the measured pressures generated from different charge shapes? The K&B expression for the incident or reflected pressure Y calculation is:

$$Y = 10^{(C_0 + C_1 \times U + C_2 \times U^2 + \cdots + C_N \times U^N)} \text{ with } U = K_0 + K_1 \times \log Z, \tag{1}$$

where Z is scaled distance, and K_0, K_1, C_0, C_1, $C_2 \dots C_N$ are Kingery and Bulmash polynomial constants, which can be found in Kingery and Bulmash [8].

5 RESULTS AND DISCUSSION

The experimental results were analysed and compared to the results obtained by K&B equation and numerical simulations in ANSYS Autodyn.

5.1 Experimental results

Four detonations were performed for each charge shape and four incident, and two reflected, pressures were measured during each detonation.

5.1.1 Incident pressure

The diagrams of the distribution of incident pressure versus time are shown in Fig. 5, for all measurement sensors. For the spherical shape, an equal pressure distribution is expected due to the spherical propagation of the blast wave. Differences in the diagrams are possible because of the imperfect sphere shape due to hand moulding. The detonator placed from below results in the pressure diagrams being more uniform over time because the position of the detonator has the same effect on explosive material in all four directions. All measurements have the duration of the positive phase, up to 1.5 ms. A horizontally oriented cylindrical charge is expected to have higher pressures perpendicular to the cylinder but a vertically oriented cylinder will have more uniform pressures in all directions and similar arrival times.

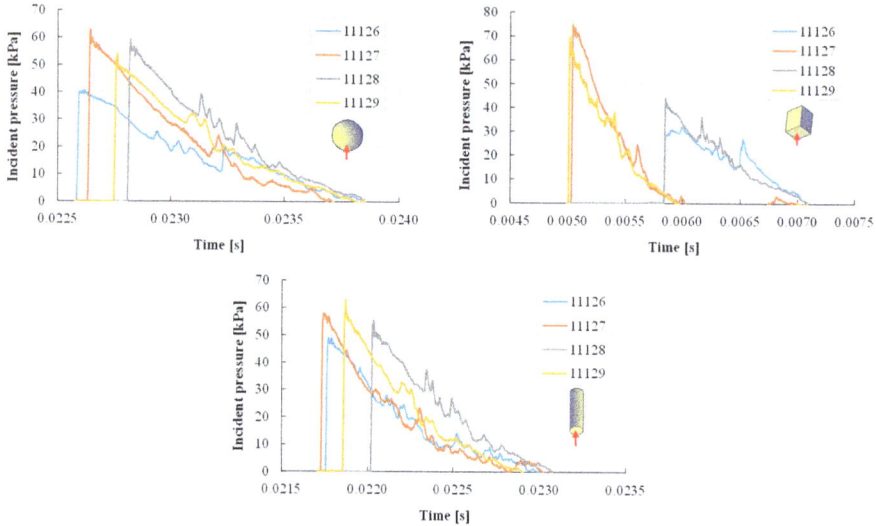

Figure 5: Incident pressure–time diagrams measured for all three charge shapes.

A rectangular charge produced the highest incident pressure of 75.02 kPa. Cylindrical and spherical explosive charges exhibit lower pressure values, 63.06 kPa and 62.89 kPa, respectively. The difference between the maximum measured pressures reached 16% but was within acceptable limits.

Table 2 presents all of the measured pressures. The arithmetic means, the standard deviation, and the coefficient of variation (CoV) were calculated for each sensor. CoV indicates a high reliability of experimental results because deviations are within the allowable limits, i.e. less than 30%.

Table 2: Measured incident pressures (kPa) in all detonations.

	Spherical				Rectangular				Cylindrical			
Detonation	11126	11127	11128	11129	11126	11127	11128	11129	11126	11127	11128	11129
1	48.59	52.63	57.23	51.75	30.22	64.83	28.28	60.64	41.95	53.75	**58.26**	55.32
2	46.79	57.85	45.33	53.89	**27.88**	60.85	34.17	65.23	49.09	58.12	55.27	**63.06**
3	40.60	**62.89**	59.31	54.24	29.06	68.10	33.46	66.70	51.21	54.27	54.38	55.54
4	48.74	55.07	53.31	53.76	31.98	**75.02**	43.93	69.61	51.00	57.20	54.66	52.95
$\bar{x}$	**46.18**	**57.11**	**53.80**	**53.41**	**29.79**	**67.20**	**34.96**	**65.55**	**48.31**	**55.84**	**55.64**	**56.72**
σ	3.83	4.40	6.17	1.13	1.75	6.00	6.53	3.74	4.35	2.15	1.78	4.39
CoV	8%	8%	11%	2%	6%	9%	19%	6%	9%	4%	3%	8%

Note: $\bar{x}$ = arithmetic mean, σ = standard deviation, CoV = coefficient of variation.

Pressure–time diagrams measured on sensor 11128 for all charge shapes are shown in Fig. 6(a). The rectangular charge gives the lowest pressures on this sensor, while the minimum differences are between the sphere and cylinder. Experimental measurements did not contain the arrival time and, because of that, when comparing results, the maximum pressures were set at the same instance.

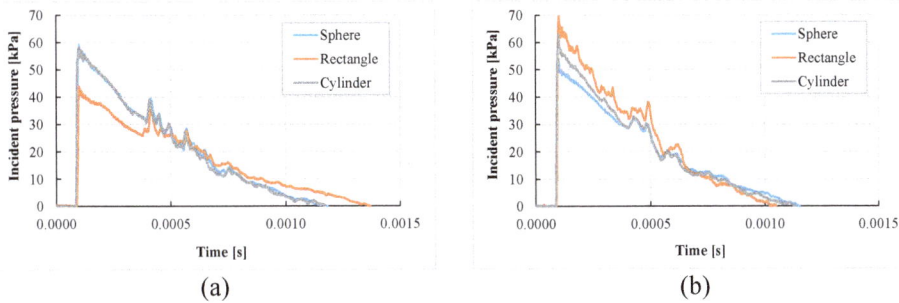

Figure 6:   Measured incident pressure–time diagrams. (a) Sensor 11128; and (b) Sensor 11129.

The diagrams in Fig. 6(b) show pressures for the lateral sensor 11129, perpendicular to the direction of charge. The records show a uniform appearance of the pressure diagram for all three charge shapes. The difference lies in the peak pressure values; the rectangular shape shows a slightly pronounced pressure peak, while the sphere shows the lowest peak pressure.

5.1.2 Reflected pressure
Table 3 presents the measured reflected pressures. The most uniform measurements are for the sphere (coefficient of variation (CoV) 7%), then the rectangle (from 9% to 13%), and the cylinder (from 5% to 20%). The CoV of 20% for sensor 5771 indicates significant differences in measurements but is still within acceptable limits.

Table 3: Measured reflected pressures (kPa).

	Spherical		Rectangular		Cylindrical	
	5770	5771	5770	5771	5770	5771
Bottom	188.84	258.47	246.62	308.46	188.31	230.44
	176.11	277.59	189.76	277.17	205.06	314.72
	192.86	297.82	244.35	311.17	193.48	200.19
	210.44	298.66	208.16	258.39	182.11	282.38
$\bar{x}$	192.06	283.13	222.22	288.80	192.24	256.93
σ	14.18	19.11	27.91	25.47	9.73	51.34
CoV	7%	7%	13%	9%	5%	20%

Note: $\bar{x}$ = arithmetic mean, σ = standard deviation, CoV = coefficient of variation.

The diagram in Fig. 7 shows the reflected pressures measured on sensor 5770 during the detonation of all three shapes. The diagrams are uniform, which indicates the accuracy of the measurements. For the detonator placed vertically, measurements show the highest reflected pressures for the rectangle.

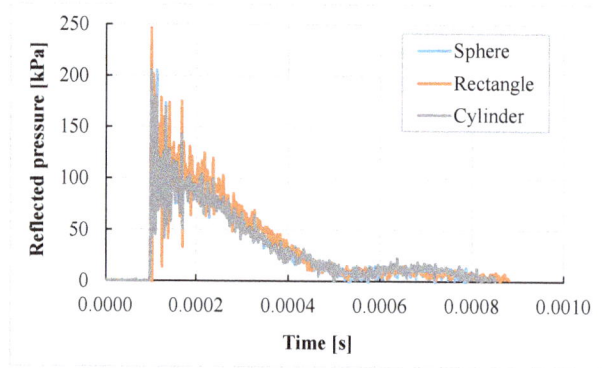

Figure 7: Reflected pressure–time diagrams measured by 5770.

5.2 Numerical results

The sphere was modelled as an axisymmetric (1D) model with a mesh size of 1 mm for air elements and is used for K&B calculated pressures verification. The base model for simulating experimental tests was a 2D planar model with a mesh size of 2.5 mm.

5.2.1 Incident pressure

Incident pressures were measured with a gauge 4 sensor (11128) for all charge shapes and are shown in Fig. 8(a). Comparing the diagrams of the incident pressures obtained numerically, the difference in arrival time of the blast wave to the sensor is also visible. There is also a difference in the measured pressure values. The smallest value of measured pressure is for the cylinder. The difference in value is almost 30 kPa, in comparison with the spherical charge.

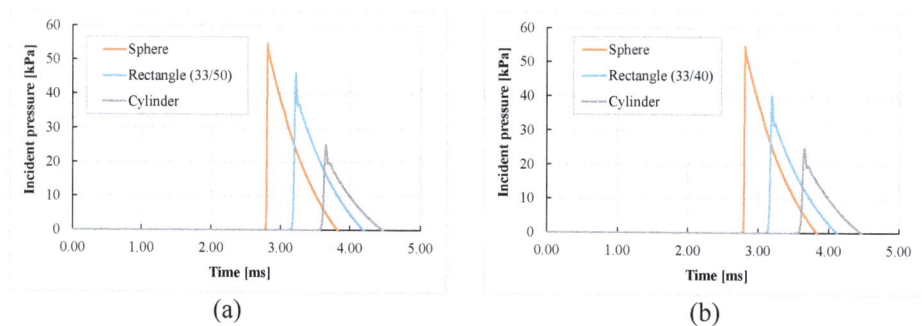

Figure 8: Numerically simulated incident pressure–time diagrams. (a) Gauge 4 (11128); and (b) Gauge 2 (11127 and 11129).

The pressures were measured at the lateral gauge point (2), perpendicular to the charge axis. Data correspond to the results of experimental measurements on sensors 11127 and 11129. The diagrams are uniform for all three shapes (Fig. 8(b)) but differ in their maximum value. The pressures measured for the sphere and cylinder are the same as for gauge 4, which was expected because of the circular charge shape and uniform blast wave propagation.

5.2.2 Reflected pressure

The reflected pressure was measured at point 3, shown in Fig. 4. The reflected pressure diagrams follow the trend of the incident pressure diagrams but are of greater intensity (Fig. 9). A vertically placed detonator results in the lowest reflected pressures for the cylinder and the highest for the spherical charge detonation.

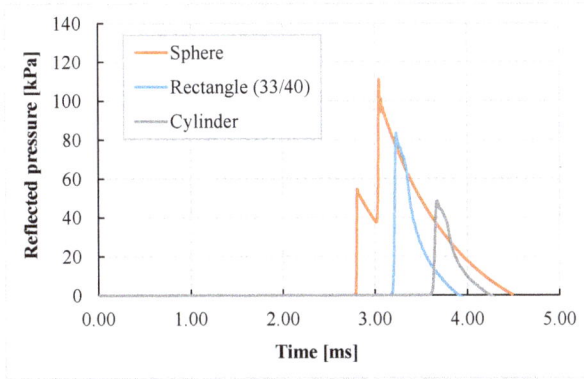

Figure 9: Reflected pressure–time diagrams measured on the gauge 3 (5770 and 5771).

5.3 Kingery Bulmash expression

Eqn (1) provides pressure values for charges placed in the air (A) (sphere) and on the surface (S) (hemisphere). Table 4 lists the K&B values. Incorporating the shape as a parameter in K&B expressions is not applicable. The pressures obtained for the sphere served in the comparison of all detonation scenarios.

Table 4: Values obtained via Kingery and Bulmash expressions.

Parameters			Kingery and Bulmash				
W (kg)	R (m)	TNT equivalent	Position	Incident (kPa)	Reflected (kPa)	t_a (ms)	t_d (ms)
0.100	1.857	1.3	Air (A)	54.90	133.50	2.8	1.50
0.100	1.857	1.3	Surface (S)	76.90	199.00	2.5	1.65

Note: t_a = arrival time, t_d = positive phase duration.

5.4 Result comparison

The arithmetic mean of the experimentally determined (incident and reflected) pressures for different charge shapes and the pressure values determined by the Kingery and Bulmash expression, are shown in Fig. 10.

A comparison of experimental and K&B results shows that K&B overestimate the pressures (Fig. 10(a)). Significantly higher pressures were recorded by the lateral sensors (11127 and 11129) for all charge shapes, especially for a rectangular charge. The measured results coincide with K&B results when the charge is placed in the air, which is to be expected. According to Fig. 10(b), it can be seen that all of the reflected pressures measured experimentally were closer to the value for the surface burst obtained via K&B.

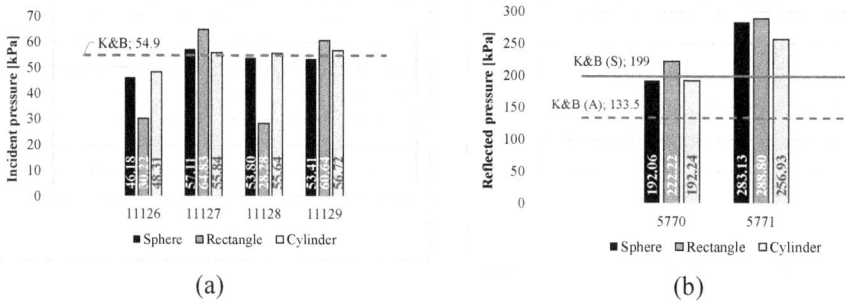

(a)   (b)

Figure 10:   Arithmetic mean of measured pressures. (a) Incident pressures; and (b) Reflected pressures.

A sphere modelled in 1D axisymmetry gives better matches with K&B than modelling in 2D because of a finer mesh (1 mm) which provides more accurate results. For the sphere, the incident pressures obtained were the same as those found through the K&B expression, while the rectangular and cylindrical charges produced lower pressures (Fig. 11). The comparison indicates the difference in pressures for different charge shapes.

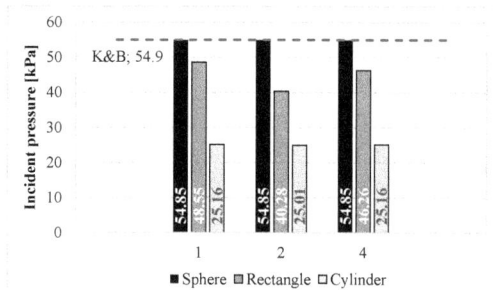

Figure 11:   Numerically obtained incident pressures.

Comparing the numerically obtained reflected pressures with K&B, they are lower than the pressures obtained via the K&B expression for a charge in free-air. In the experimental results, the measured pressures were higher than those obtained for K&B for a surface charge.

A comparison of the numerically determined pressures with the arithmetic mean of the experimentally determined ones shows a good agreement of the results, considering the spherical charge. For the rectangular shape, experimentally similar pressures were measured with sensors 11126 and 11128, which are lower than the numerically determined pressures at these locations. With sensors 11127 and 11129, pressures twice as high as those measured with sensors 11126 and 11128 were measured, and these pressures were higher than the numerically determined pressures. It is believed that the differences are due to the fact that the numerical model was created in a 2D simulation and the third dimension of the charge shape, which also affects the measured pressure, was neglected. This problem also occurred with the cylinder because the influence of the height of the cylinder on the measured pressure is neglected in the numerical simulation. After detonation of the cylinder, uniform pressures were measured at all sensors, which is to be expected because the cylinder is vertical, but the experimental pressure measurements were more than twice the numerically determined pressures; the results are shown in Fig. 12.

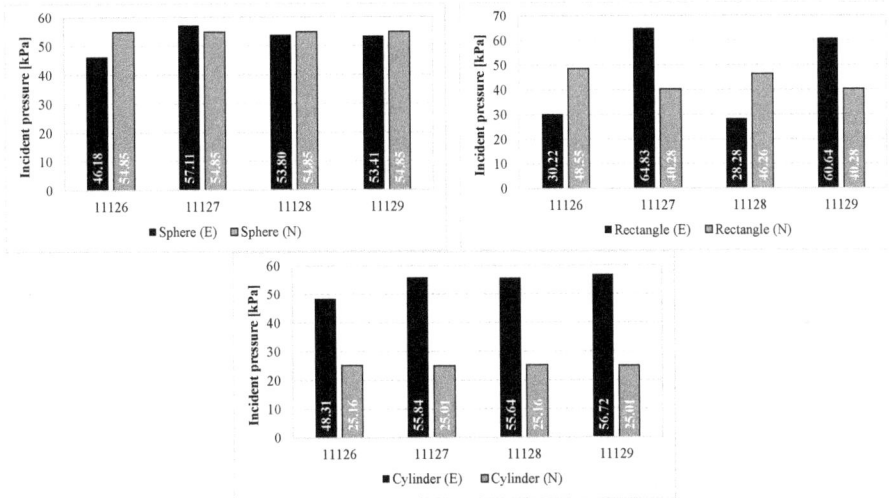

Figure 12: Comparison of experimentally measured and numerically determined pressures.

6 CONCLUSION

The Kingery and Bulmash equation predicts the pressures well for the spherical charge. The deviations from the experiments and the expression are minimal. Numerical simulations can also give identical pressures using a simple 1D model with a mesh size of 1 mm. When rectangular and cylindrical explosive charges are considered, the comparison yields significant discrepancies. For vertically oriented explosives, K&B estimates pressures well for these two shapes. It is recommended that the influence of shape and orientation be further investigated by increasing the weight of the charge and varying the standoff distance. In numerical simulations, K&B overestimates the pressures for the rectangle and cylinder. The numerically determined reflected pressures are much lower than the pressures measured experimentally and calculated by K&B. Due to the limited capacity of computer hardware, only 1D and 2D models are considered in this paper. This should be extended to 3D simulations to include all three charge dimensions, since 2D simulations are spatially limited.

ACKNOWLEDGEMENTS

This paper was supported, in part, by the Croatian Science Foundation (HRZZ) under the project (UIP-2017-05-7041) "Blast Load Capacity of Highway Bridge Columns", and support for this research is gratefully acknowledged. The authors would also like to thank the Croatian Army for providing all the necessary support in conducting blast measurements in the form of recommendations, access to the training ground, and securing the explosive material.

REFERENCES

[1]   Wu, C., Fattori, G., Whittaker, A. & Oehlers, D.J., Investigation of air-blast effects from spherical-and cylindrical-shaped charges. *International Journal of Protective Structures*, **1**, pp. 345–362, 2010.

[2] Sherkar, P., Shin, J., Whittaker, A. & Aref, A., Influence of charge shape and point of detonation on blast-resistant design. *Journal of Structural Engineering*, **142**, 04015109, 2016.

[3] Artero-Guerrero, J., Pernas-Sánchez, J. & Teixeira-Dias, F., Blast wave dynamics: The influence of the shape of the explosive. *Journal of Hazardous Materials*, **331**, pp. 189–199, 2017

[4] Langran-Wheeler, C., Tyas, A., Rigby, S., Stephens, C., Clarke, S. & Warren, J., Characterisation of reflected blast loads in the very-near field from non-spherical explosive charges. *Proceedings of the 17th International Symposium for the Interaction of Munitions with Structures (ISIEMS17)*, 2017.

[5] Langran-Wheeler, C., Tyas, A., Rigby, S., Stephens, C., Clarke, S. & Walker, R., Reflected blast loads from long cylinders in the near-field. *Proceedings of the 18th International Symposium for the Interaction of Munitions with Structures (ISIEMS18)*, Panama City, FL, 2019.

[6] Hu, Y., Chen, L., Fang, Q. & Xiang, H., Blast loading model of the RC column under close-in explosion induced by the double-end-initiation explosive cylinder. *Engineering Structures*, **175**, pp. 304–321, 2018.

[7] Xiao, W., Andrae, M. & Gebbeken, N., Influence of charge shape and point of detonation of high explosive cylinders detonated on ground surface on blast-resistant design. *International Journal of Mechanical Sciences*, **181**, 105697, 2020.

[8] Kingery, C.N. & Bulmash, G., Airblast parameters from TNT spherical air burst and hemispherical surface burst. US Army Armament and Development Center, Ballistic Research Laboratory, 1984.

[9] Ngo, T., Mendis, P., Gupta, A. & Ramsay, J., Blast loading and blast effects on structures: An overview. *Electronic Journal of Structural Engineering*, **7**, pp. 76–91, 2007.

[10] Lukić, S., Draganić, H., Gazić, G. & Radić, I., Statistical analysis of blast wave decay coefficient and maximum pressure based on experimental results. *WIT Transaction on The Built Environment*, vol. 198, WIT Press: Southampton and Boston, pp. 65–78, 2020.

[11] ANSYS I, ANSYS Autodyn user manual (Release 15.0). ANSYS, 2013.

[12] Kohnke, P., ANSYS Theory Manual, Release 12.0. ANSYS Inc., 2009.

[13] Draganić, H. & Varevac, D., Numerical simulation of effect of explosive action on overpasses. *Građevinar*, **69**, pp. 437–451, 2017.

[14] Hyde, D.W., Microcomputer programs CONWEP and FUNPRO, Applications of TM 5-855-1, "Fundamentals of Protective Design for Conventional Weapons" (User's Guide). Army Engineer Waterways Experiment Station, Vicksburg MS, Structures Laboratory, 1988.

NEW PREDICTIVE MODELS FOR THE BALLISTIC LIMIT OF SPACECRAFT SANDWICH PANELS SUBJECTED TO HYPERVELOCITY IMPACT

ALEKSANDR CHERNIAEV & RILEY CARRIERE
Department of Mechanical, Automotive and Materials Engineering, University of Windsor, Canada

ABSTRACT

Cell size, foil thickness, and the material of the core, influence the ballistic performance of honeycomb-core sandwich panels (HCSP) in the case of hypervelocity impact (HVI) by orbital debris. Two predictive models that account for this influence have been developed in this study: a dedicated ballistic limit equation (BLE) and an artificial neural network (ANN) trained to predict the outcomes of HVI on HCSP. The BLE is a modified version of the Whipple shield BLE and demonstrated excellent accuracy in predicting the ballistic limits of HCSP, when tested against a new set of simulation data, with the discrepancy ranging from 1.13% to 5.58% only. The ANN was developed using MATLAB's Deep Learning Toolbox framework and was trained utilizing the same HCSP HVI database as that employed for the BLE fitting and demonstrated a very good predictive accuracy, when tested against a set of simulation data not previously used in the training of the network, with the discrepancy ranging from 0.67% to 7.27%.
Keywords: orbital debris shielding, honeycomb-core sandwich panels, hypervelocity impact, ballistic limit equation, artificial neural network.

1 INTRODUCTION

For unmanned spacecraft, it is often possible to use pre-existing structural components as orbital debris shielding, thus designing multifunctional structures and enabling additional weight savings [1]. In a typical satellite design, most impact-sensitive equipment is situated in the enclosure of the structural honeycomb-core sandwich panels (HCSP). Being the most commonly used elements of satellite structures, these panels form the satellite's shape and are primarily designed to resist launch loads and provide attachment points for satellite subsystems [2]. With low additional weight penalties, their intrinsic ballistic performance can often be upgraded to the level required for orbital debris protection [3].

Assessing the orbital debris impact survivability of unmanned satellites requires reliable predictive models for honeycomb-core sandwich panels, capable of accounting for various impact conditions and panel design parameters. To make it applicable for HCSP and account for the presence of honeycomb core, Lathrop and Sennett [4] proposed modifying the well-known Whipple shield ballistic limit equation (eqn (1)) (normal incidence impact-only version):

$$D_{cr} = 3.918 \cdot \sqrt[3]{\frac{\bar{S}}{\rho_p \sqrt[3]{\rho_b}} \cdot \left(\frac{t_{FC}}{v_p}\right)^2 \cdot \left(\frac{\sigma_{Y,FC}}{70}\right)}, \tag{1}$$

by replacing standoff distance $\bar{S}$ in it by either the product of twice the honeycomb cell size (A_{cell}) or by the core depth (t_{HC}), whichever is less: $\bar{S} = \min(2 \cdot A_{cell}, t_{HC})$. Here ρ_p and ρ_b are the projectile and front facesheet ("bumper") densities in g/cm^3; t_{FC} – thickness of the rear facesheet in mm; v_p – projectile speed in km/s; $\sigma_{Y,FC}$ – facesheet yield strength in ksi. This approach, however, is considered to be a "rough estimate" [5] and does not include other influential parameters, such as foil thickness and material of the core.

WIT Transactions on The Built Environment, Vol 209, © 2022 WIT Press
www.witpress.com, ISSN 1743-3509 (on-line)
doi:10.2495/HPSU220091

In this study, two distinct methods were used to construct honeycomb core parameters sensitive predictive models: a conventional approach based on the development of HCSP-specific ballistic limit equation (BLE), and an approach based on the development and training of an artificial neural network (ANN). The developed predictive models are focused on the most conservative scenario of hypervelocity impact (HVI) at the normal incidence and limited to aluminum HCSP.

2 HVI DATABASE

Implementation of both of these methods relied on a database for HVI on HCSP, which was extracted from Carriere and Cherniaev [6] and included 56 entries. Different panel configurations and their respective ballistic limits, derived from this database, are summarized in Table 1. Here ballistic limit is understood as particle size on threshold of the critical damage mode – perforation/no perforation of the rear facesheet in case sandwich panel targets.

Table 1: Ballistic limits of HCSP configurations considered in the development of new BLE and ANN.

Projectile		Facesheets		Honeycomb		Ballistic limit
Speed, km/s	Material	Material	Thickness, mm	Grade*	Depth, mm	D_{cr}, mm
6.80	Al2017-T4	Al6061-T6	0.41	1/8-5052-0.003	12.7	0.90
6.75	Al2017-T4	Al7075-T6	1.60	3/16-5056-0.001	50.0	1.71
7.00	Al2017-T4	Al6061-T6	1.60	1/8-5052-0.001	25.0	1.70
7.00	Al2017-T4	Al6061-T6	1.60	3/16-5052-0.001	25.0	2.50
7.00	Al2017-T4	Al6061-T6	1.60	1/4-5052-0.001	25.0	2.50
7.00	Al2017-T4	Al6061-T6	1.60	1/8-5052-0.003	25.0	1.50
7.00	Al2017-T4	Al6061-T6	1.60	3/16-5052-0.003	25.0	1.90
7.00	Al2017-T4	Al6061-T6	1.60	1/4-5052-0.003	25.0	2.10
7.00	Al2017-T4	Al6061-T6	1.00	1/8-5052-0.001	50.0	1.10
7.00	Al2017-T4	Al6061-T6	1.00	3/16-5052-0.001	50.0	1.30
7.00	Al2017-T4	Al6061-T6	1.00	1/4-5052-0.001	50.0	1.50
7.00	Al2017-T4	Al6061-T6	1.00	1/8-5052-0.003	50.0	1.10
7.00	Al2017-T4	Al6061-T6	1.00	3/16-5052-0.003	50.0	1.10
7.00	Al2017-T4	Al6061-T6	1.00	1/4-5052-0.003	50.0	1.30
7.00	Al2017-T4	Al6061-T6	1.60	1/8-5052-0.001	50.0	1.50
7.00	Al2017-T4	Al6061-T6	1.60	3/16-5052-0.001	50.0	1.90
7.00	Al2017-T4	Al6061-T6	1.60	1/4-5052-0.001	50.0	2.30
7.00	Al2017-T4	Al6061-T6	1.60	1/8-5052-0.003	50.0	1.50
7.00	Al2017-T4	Al6061-T6	1.60	3/16-5052-0.003	50.0	1.70
7.00	Al2017-T4	Al6061-T6	1.60	1/4-5052-0.003	50.0	2.10

* Honeycomb grade: cell size (in.) – honeycomb material – foil thickness (in.). For example, 1/8-5052-0.003 stands for a honeycomb with 3.18 mm (1/8 in.) cells made of Al5052 and having a foil thickness of 0.076 mm (0.003 in.).

3 BALLISTIC LIMIT EQUATION

The new BLE for HVI on HCSP proposed in this study is a modification of the Whipple shield BLE, which does not alter the general expression provided by eqn (1), however the expression for $\bar{S}$ in our BLE was supplemented by additional terms, such that:

$$\bar{S} = K \cdot A_{cell} \cdot \left(\frac{t_{HC}}{t_{FC}+\alpha}\right)^{\beta} \cdot \left(\frac{t_{HC}}{t_{foil}}\right)^{\gamma} \cdot \left(\frac{30}{\sigma_{Y,HC}}\right)^{\delta}. \tag{2}$$

Here t_{HC} = honeycomb depth in mm; t_{FC} = thickness of a facesheet in mm; t_{foil} = thickness of the honeycomb foil in mm; $\sigma_{Y,HC}$ = yield strength of the honeycomb material in ksi (e.g., 30 ksi for Al5052 and 50 ksi for Al5056 honeycomb); and K, α, β, γ, δ are parameters with the values given in Table 2.

Table 2: Parameters of the new HCSP BLE.

BLE parameter	K	α	β	γ	δ
Value	2.63	1.893	−0.804	0.304	1.915

The new BLE fit factors presented in Table 2 were determined by minimizing the discrepancy (expressed in terms of the sum of squared errors) between the BLE predictions and the HVI database data provided in Table 1. The goodness-of-fit diagram for the BLE proposed in this study is shown in Fig. 1.

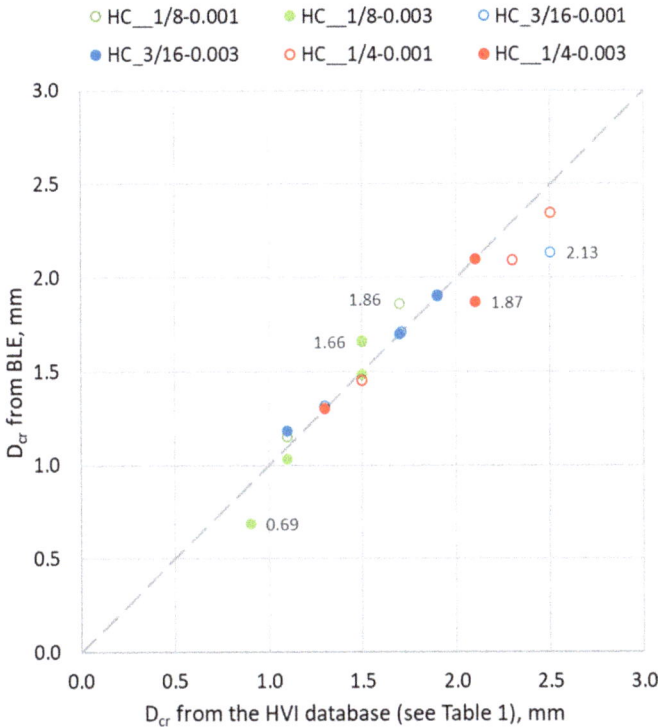

Figure 1: Goodness of fit diagram for the new BLE.

4 ARTIFICIAL NEURAL NETWORK

In this study, MATLAB's Deep Learning Toolbox was used to develop an ANN capable of predicting perforating/non-perforating outcomes of HVI of all-aluminum HCSP structures and projectiles at normal incidence. A binary output classification scheme was established with the pass "non-perforating" and fail "perforating" classes set. A perforating case is defined as being when the fragments fully penetrate through the rear facesheet.

Input parameters for the ANN included projectile speed and diameter, facesheet material and thickness, honeycomb depth, material, foil thickness and cell size. The developed ANN featured root mean squared propagation activation function, node sizing of 3 and a single hidden layer.

Training sets were established using a hold-out validation scheme, using 80% of the database, for the ANN to learn and develop relations. Once developed, the remaining 20% of the database located in the testing set, were used as "true" prediction scenarios, allowing for analysis of the ANN's predictive accuracy against known outcomes in the database. The accuracy of ANN predictions was estimated as 85%.

5 VERIFICATION OF BLE AND ANN

Additional datapoints found in Carriere and Cherniaev [6] that have not been used in either BLE fitting or ANN training and, thus, were "unfamiliar" to both, were employed to conduct verification of the developed predictive models. It should be noted that data in Carriere and Cherniaev [6] was obtained by running a detailed numerical model of HVI on HCSP.

Table 3 compares the ballistic limit predictions of the new BLE, ANN and the data from Carriere and Cherniaev [6]. As can be deduced from the table, in all cases, the BLE demonstrated an excellent correlation with the predictions of the sophisticated numerical model, with the discrepancy ranging from 1.13% to 5.58% only. Critical projectile diameter estimations by the ANN closely resembled the simulation ballistic limits: the difference between simulation and ANN predictions ranged between 0.67% and 7.27%.

Table 3: Verification of BLE and ANN predictions.

Projectile		Facesheets			Honeycomb		Ballistic limit		
Speed (km/s)	Material	Material	Thickness (mm)	Grade*	Depth (mm)	D_{cr} (mm)			
						Ref. [6]	BLE	ANN	
7.00	Al2017-T4	Al6061-T6	1.30	1/8-5052-0.001	25.0	1.50	1.58	1.53
7.00	Al2017-T4	Al6061-T6	1.60	3/16-5052-0.003	38.0	1.70	1.78	1.76
7.00	Al2017-T4	Al6061-T6	1.00	5/32-5052-0.002	50.0	1.10	1.16	1.18
7.00	Al2017-T4	Al6061-T6	1.30	5/32-5052-0.002	38.0	1.50	1.48	1.51
7.00	Al2017-T4	Al7075-T6	1.00	1/4-5056-0.001	50.0	1.30	1.26	1.26

6 CONCLUSIONS

The new ballistic limit equation is based on the Whipple shield BLE, in which the standoff distance between the facesheets was replaced by a function of the honeycomb cell size, foil thickness, and yield strength of the HC material. The corresponding fit factors were determined by minimizing the sum of squared errors between the BLE predictions and the results of HVI tests listed in the database. The BLE was then tested against a new set of simulation data and demonstrated an excellent predictive accuracy, with the discrepancy ranging from 1.13% to 5.58% only.

The artificial neural network was developed using MATLAB's Deep Learning Toolbox framework and was trained utilizing the same HCSP HVI database as was employed for the BLE fitting. The developed ANN utilized the root mean square propagation activation function and one hidden layer with three nodes. The ANN demonstrated a very good predictive accuracy, when tested against a set of simulation data not previously used in the training of the network, with the discrepancy ranging from 0.67% to 7.27%.

ACKNOWLEDGEMENT

This work was financially supported by the Natural Sciences and Engineering Research Council of Canada through Discovery Grant No. RGPIN-2019-03922 "Orbital debris impact survivability models for robotic satellites".

REFERENCES

[1]   Adams, D.O. et al., Multi-functional sandwich composites for spacecraft applications: An initial assessment. NASA/CR-2007-214880, 2007.

[2]   Bylander, L.A., Carlström, O.H., Christenson, T.S.R. & Olsson, F.G.A., Modular design concept for small satellites. *Smaller Satellites: Bigger Business?*, Springer: Netherlands, pp. 357–358, 2002.

[3]   Cherniaev, A. & Telichev, I., Weight-efficiency of conventional shielding systems in protecting unmanned spacecraft from orbital debris. *Journal of Spacecraft and Rockets*, **54**(1), pp. 75–89, 2016.

[4]   Lathrop, B. & Sennett, R., The effects of hypervelocity impact on honeycomb structures. *9th Structural Dynamics and Materials Conference*, American Institute of Aeronautics and Astronautics, 1968.

[5]   Christiansen, E.L. et al., Handbook for Designing MMOD Protection, NASA JSC-64399, Version A, JSC-17763, 2009.

[6]   Carriere, R. & Cherniaev, A., Honeycomb parameter-sensitive predictive models for ballistic limit of spacecraft sandwich panels subjected to hypervelocity impact at normal incidence. *Journal of Aerospace Engineering*, **35**(4), 2022. DOI: 10.1061/(ASCE)AS.1943-5525.0001436.

ASSESSMENT OF COUPLED LAGRANGIAN–EULERIAN FINITE ELEMENT SIMULATIONS TO MODEL SUCTION FORCES DURING HYDRODYNAMIC IMPACTS

MATHIEU GORON[1,2], BERTRAND LANGRAND[1,3], THOMAS FOUREST[1],
NICOLAS JACQUES[2] & ALAN TASSIN[4]
[1]DMAS, ONERA, France
[2]ENSTA Bretagne, France
[3]University Polytechnique Hauts-de-France, France
[4]IFREMER, France

ABSTRACT

During the emergency landing of an aircraft on water, the structure may experience critical forces and could eventually fail. The appropriate design of the structure should minimize the risk of occupant injuries. The recent progress in computation capabilities led to the increased use of numerical simulations in the certification process of aircraft. A specific challenge concerns the modelisation of suction forces that develop near the aircraft tail, where the first contact with water occurs. This phenomenon is due to the high horizontal velocity of the structure at impact and the longitudinal curvature of the fuselage. It can affect the overall aircraft kinematics during ditching. In this work, as an effort to improve aircraft ditching simulations and to assess the capabilities of numerical models to describe suction forces, the simple test case of the wedge water entry and subsequent exit is considered. Numerical simulations with the Eulerian formulation for the fluid and the Lagrangian formulation for the structure are used. The method used for the fluid–structure interaction is based on an immersed contact interface with penalty forces. The present work focuses on impact and suction forces modelling. Results show a satisfying capacity of the numerical approach to model negative hydrodynamic force (suction).

Keywords: finite elements, fluid–structure interaction, Eulerian–Lagrangian coupling, water impact and exit, hydrodynamic forces, suction.

1 INTRODUCTION

A hydrodynamic impact is a contact between a structure and a fluid in relative motion. The study of this phenomenon is motivated by various applications such as hull slamming, spacecraft (capsule), aircraft and rotorcraft emergency water landing (ditching). The hydrodynamic forces caused by water impacts are critical and considered during the sizing and certification of structures exposed to this kind of event.

The physical phenomena encountered during hydrodynamic impacts are well known for "ideal" impact conditions: vertical impact, simple geometry, rigid structure, fluid initially at rest, impact velocity large enough to neglect gravity, etc. Analytical approaches, mainly based on von Karman's [1] or Wagner's [2] theories, have been developed to analyse the loading applied to the structure. Numerous articles can be found in the literature, addressing the problem of hydrodynamic impact of simple geometries experimentally (wedges [3]–[5], circular cylinders [6], [7], cones [8], flat plates [9], spheres [10]) and numerically [11]–[14]. The recent advances in numerical simulation have gradually allowed the study of more complex phenomena (considering certain non-linearities related to the fluid–structure interaction, complex geometries, etc.), representative of realistic industrial applications. The finite element (FE) method, with a Lagrangian framework, is usually used to model the structure. The fluid behaviour can be described by various methods: Lagrangian [15],

WIT Transactions on The Built Environment, Vol 209, © 2022 WIT Press
www.witpress.com, ISSN 1743-3509 (on-line)
doi:10.2495/HPSU220101

Eulerian [16], [17], arbitrary Lagrangian–Eulerian (ALE) [18], or mesh-free methods such as smoothed particle hydrodynamics (SPH) [16].

The physical phenomena encountered during realistic industrial applications are more complex. For instance, during high-velocity oblique impacts suction, cavitation, ventilation, aeration or flow separation phenomena can appear and incapacitate the existing analytical and numerical approaches. Using these approaches in a sizing or certification context is therefore difficult at the moment as they require further developments.

In the present work, the explicit FE solver Radioss is used. A coupled Eulerian–Lagrangian (CEL) approach is considered to model ditching or water impact problems. On the one hand, the structure is described by a Lagrangian approach. On the other hand, the fluid behaviour is described by an Eulerian approach. A strong (two-way) coupling between the fluid and structural solutions is used because the interdependence of the fluid and structure models is important [19], [20].

The objective of this work is to assess: (i) the ability of this numerical method to model a hydrodynamic impact; (ii) the sensibility of the method to the numerical parameters; and (iii) the ability of the method to model some hydrodynamic phenomena taking place during a realistic industrial application, namely the suction. To do so, a simple test case is considered. It consists in the high-velocity and vertical impact of a wedge, based on the experiments presented in Richard [5]. Numerical results are compared to experimental results regarding hydrodynamic force and pressure measured on the structure.

This article is organised as follows. Section 2 provides a brief description of the numerical method. In Section 3, the effects of the method numerical parameters on the impact forces are discussed and the numerical and experimental results are compared. Section 4 introduces the case of the water entry and subsequent exit of a wedge. The capacity of the numerical method to model suction is thereby discussed. Finally, conclusions are drawn and orientations for future research are discussed in Section 5.

2 NUMERICAL METHOD

This section presents the method used to model the vertical impact of a wedge with the considered FE explicit solver, based on the experiments presented in Richard [5].

2.1 Structure modelling

The dimensions of the wedge are synthesised in Table 1 and corresponds to the ones described in Richard [5]. The wedge is modelled by two surfaces using 4-node shell FEs and is defined as a rigid body.

Table 1: Wedge dimensions.

Length L (mm)	Width B (mm)	Heel angle β (°)
495	320	20

2.2 Fluid modelling

The geometry of the fluid model is a rectangular parallelepiped that materializes the water and air around the structure. The air and water domains are discretised using 3D continuum 8-node FEs with 1 integration point. An Eulerian formulation is used to solve the fluid problem. The size of the fluid elements is smaller near the structure (in the impact zone) to increase the computation accuracy. The dimensions of the fluid domain are given in Fig. 1.

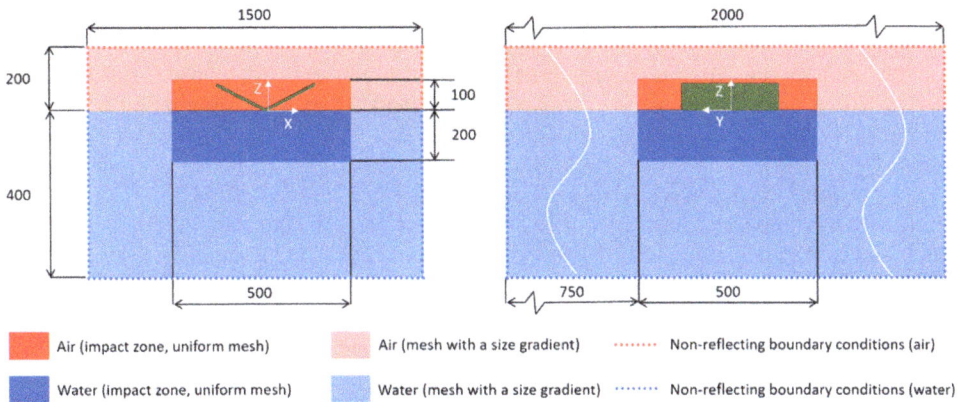

Figure 1:   Dimensions of the fluid domain (in mm). Outside the impact zone, the size of the fluid elements scales with a factor 1.2. The wedge is represented in green.

2.3 Fluid–structure interaction

The fluid–structure interaction method uses structural Lagrangian elements (master), immersed in the Eulerian fluid grid (slave nodes). The structure and the fluids are meshed independently and superposed. The coupling forces are computed using a penalty method, depending on the contact height h_c and the contact stiffness k_c. The Radioss documentation suggests defining these parameters as described in eqns (1) and (2):

$$h_{c0} = 1.5 \times l_f, \tag{1}$$

$$k_{c0} = \frac{\rho U_{max}^2 S_{el}}{h_c}, \tag{2}$$

where l_f is the size of the fluid elements in contact with the structure, ρ the water density, U_{max} the structure maximum velocity and S_{el} the mean surface of the structural elements.

2.4 Initial and boundary conditions

The initial position of the lowest point of the wedge is $h_0 = 10$ mm (Fig. 2). An initial velocity $U_{max} = -10$ m/s is applied to the wedge in the $\vec{z}$ direction. Gravity acceleration is applied to all the nodes of the model (fluid and structure): $\vec{g} = -9.81.10^{-3}\vec{z}$ g/ms². The pressure in water is set at the hydrostatic pressure at the initial time step of the computation.

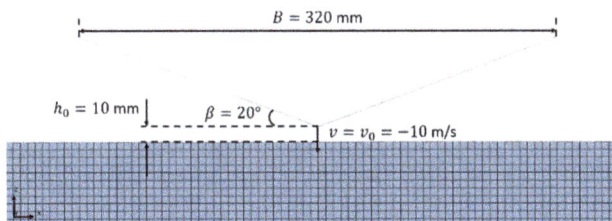

Figure 2: Initial position of the wedge above the water level.

The displacement of the mock-up is prescribed. The vertical velocity imposed to the mock-up is constant and equal to $U_{max} = -10$ m/s for the simulations presented in Section 3 and variable for the simulations presented in Section 4. Non-reflecting boundary conditions are applied on the fluid domains boundaries, based on the pressure formulation of Bayliss and Turkel [21].

3 VERTICAL AND HIGH-VELOCITY WEDGE IMPACT

Vertical wedge impact experiments have been conducted in a water tank with a high-speed shock machine as described in El Malki Alaoui et al. [22]. The tank is made of steel and is 2 m wide, 3 m long and has been filed up to 1.2 m with water during the experiments. The method used in the experiments to measure the load (hydrodynamic force, F) was based on strain gauges positioned on the piston used to move the mock-up. Pressure probes are used to measure the pressure at different locations on the wedge surface. The experimental results are presented Fig. 3 as force (slamming) and pressure coefficients (measured from a probe located on the wedge surface), obtained respectively by eqns (3) and (4).

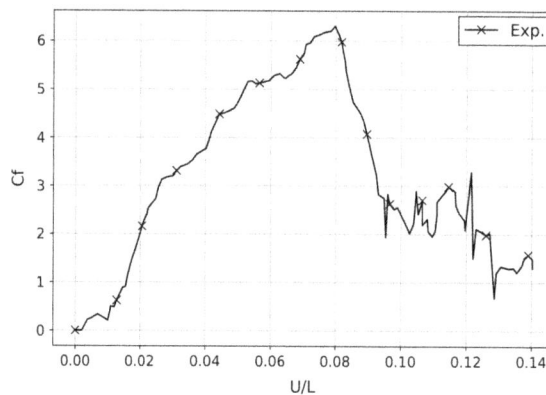

(a)

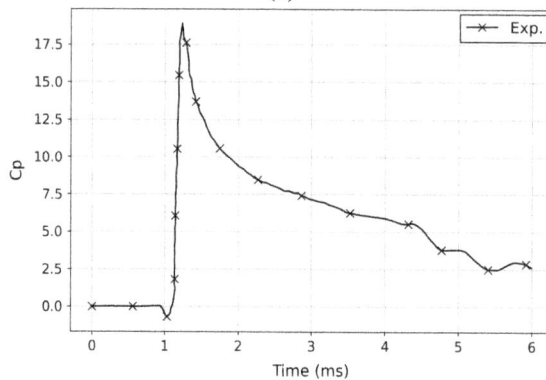

(b)

Figure 3:   Experimental vertical impact of a wedge at $U_{max} = -10$ m/s. (a) Evolution of the force coefficient depending on the vertical non-dimensional displacement $\frac{U}{L}$, with U the vertical displacement of the wedge; and (b) Time-history of the pressure coefficient [5].

$$C_f = \frac{2F}{\rho_0 U_{max}^2 S}, \quad (3)$$

$$C_p = \frac{2(P-P_0)}{\rho_0 U_{max}^2}, \quad (4)$$

where F is the hydrodynamic force, P the pressure, P_0 the atmospheric pressure, U_{max} the wedge velocity and $S = L \times B$ the wedge surface projected on the xOy plane (L and B are respectively the wedge length and width).

3.1 Investigation of different numerical parameters affecting the numerical results

In this section, the effect of different numerical parameters influencing the numerical results is presented. The following points are discussed: the effect of contact stiffness and the effect of the fluid elements size in the impact zone. The conclusions drawn are then applied to the numerical model, leading to the results presented in Section 3.2.

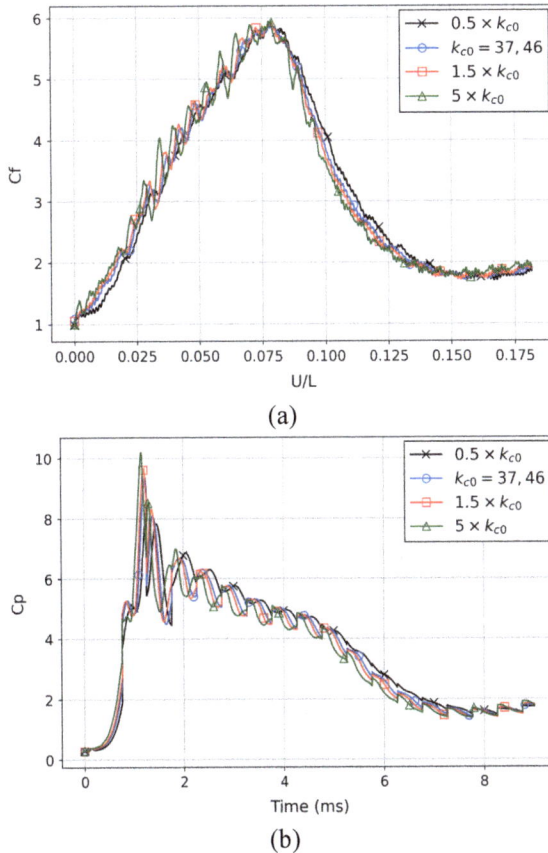

Figure 4:   Numerical vertical impact of the wedge at $U_{max} = -10$ m/s for different contact stiffnesses. (a) Evolution of the force coefficient depending on the vertical non-dimensional displacement $\frac{U}{L}$, with U the vertical displacement of the wedge; and (b) Time-history of the pressure coefficient.

3.1.1 Contact stiffness

The default contact stiffness for this case is $k_{c0} = 37.46$ (see eqn (2)). The numerical results obtained for the vertical impact of a wedge for different k_c are presented in Fig. 4. The other stiffness values considered here are $k_c = 0.5 \times k_{c0} = 18.73$, $k_c = 1.5 \times k_{c0} = 56.19$ and $k_c = 5 \times k_{c0} = 187.3$.

The influence of the contact stiffness is weak considering the variations studied (factor 10 between the smallest and the largest stiffness). The maximum force coefficient remains around $C_{f\,max} \approx 5.9$ and the trends remain similar for the different contact stiffnesses studied. For a lower contact stiffness (black curve), the early response of the model diverges from the other results and the hydrodynamic force decreases more slowly after the peak. High-frequency oscillations appear in the force response when the contact stiffness increases. The amplitude of these oscillations also increases with the value of k_c (Fig. 4(a)). The local pressure measurements are similar regardless of the contact stiffness. The amplitude of the pressure peak lightly increases with k_c (Fig. 4(b)).

3.1.2 Fluid elements size

A numerical model with two symmetry plans (xOz and yOz) is used to assess the influence of the fluid elements size on the numerical results. The different elements sizes l_f investigated here range from 5 to 2 mm. The characteristics of the different models are synthesised in Table 2.

Table 2: Vertical impact of the wedge with $U_{max} = 10$ m/s. Characteristics of the models used to assess the influence of the fluid elements size l_f on the numerical results.

l_f (mm)	Total number of fluid elements	Computation time (JJ-hh:mm:ss)
5	378,736	01:11:02
4	673,171	02:43:07
3	1,399,505	11:38:31
2.5	2,202,052	22:33:42
2	3,941,772	7-09:44:14

A decrease in the elements size in the impact zone leads to a contact height decrease (see eqn (1)) and, therefore, results in a slower rise of the force coefficient in the early impact stage. With smaller elements, the force coefficient peak is larger and the force drop is slightly faster (Fig. 5(a)). The finer the mesh size, the faster is the pressure rise and the higher is the amplitude of the pressure peak (Fig. 5(b)). The numerical results globally converge with a mesh size equal to $l_f = 2.5$ mm. However, as the model uses an explicit solver, the computational cost associated with a finer mesh size is much higher.

3.2 Comparison with experimental results

The numerical results are presented for the vertical impact of the wedge. From Section 3.1 results, the fluid elements size and contact stiffness are respectively defined as $l_f = 2.5$ mm and $k_c = 74.91$. The numerical results are compared to the experimental results from Richard [5] in terms of force and pressure coefficients in Fig. 6.

The numerical model satisfyingly predicts the experimental evolution of the hydrodynamic force and pressure. The maximum force coefficient measured numerically is

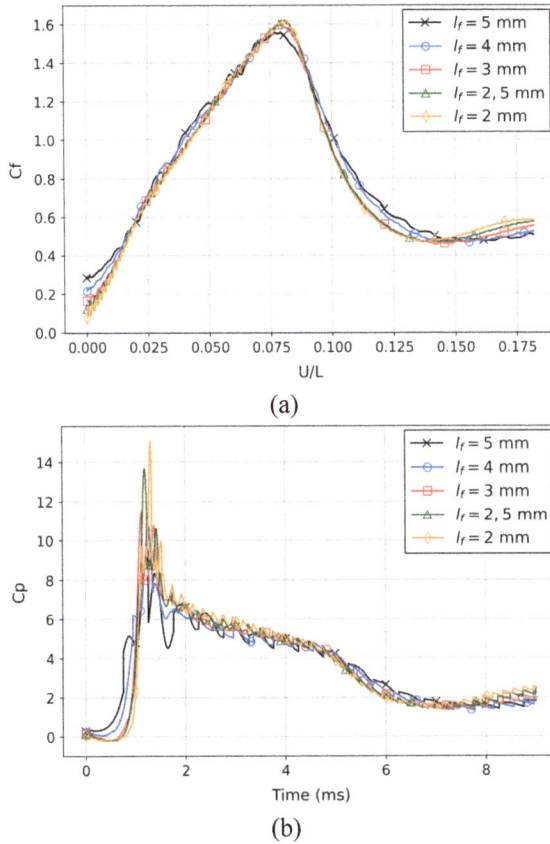

Figure 5:   Numerical vertical impact of the wedge at $U_{max} = -10$ m/s for different fluid element sizes: $l_f = [2; 2.5; 3; 4; 5]$ mm. (a) Evolution of the force coefficient as a function the vertical non-dimensional displacement $\frac{U}{L}$, with U the vertical displacement of the wedge; and (b) Time-history of the pressure coefficient.

similar to the experimental one (difference inferior to 1%). However, at the first instants of the simulation, the numerical model overestimates the hydrodynamic force level. The numerical model underestimates the decrease of the force coefficient after the force peak.

The pressure rise occurs earlier in the numerical simulation than in the experiments. Also, the numerical model underestimates the pressure coefficient peak and general level of pressure.

Some discrepancies between the numerical and experimental results can be explained by the numerical method used in the present work. Indeed, fluid–structure interaction forces appear when a fluid node is detected within the contact height of the structure. The results at the first instants of the simulation are related to the detection of fluid nodes within the contact height of the structure, before the physical contact between the wedge and the water.

There are some discrepancies between the numerical and experimental pressure evolutions. Numerically the pressure is measured at a normal distance from the wedge surface

(a)

(b)

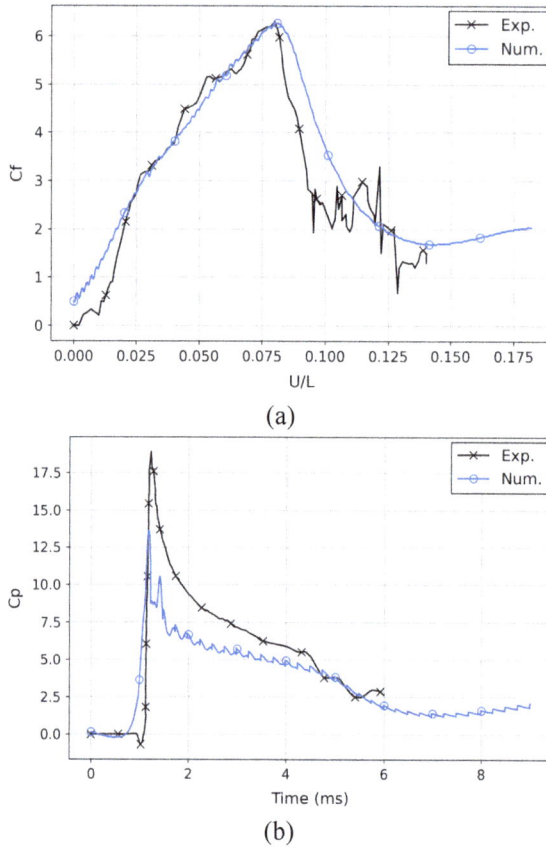

Figure 6:   Comparison of the experimental [5] and numerical results for the vertical impact of the wedge at $U_{max} = -10$ m/s. (a) Evolution of the force coefficient as a function of the vertical non-dimensional displacement $\frac{U}{L}$, with U the vertical displacement of the wedge; and (b) Time-history of the pressure coefficient.

slightly larger than the contact height ($dist = \frac{4}{3}h_{c0}$), as illustrated in Fig. 8. Experimentally the pressure probe is located on the wedge surface. The pressure peak is a spatially and temporally localised phenomenon. This explains why the present numerical method underestimates the pressure levels.

4 HIGH-VELOCITY WATER ENTRY AND SUBSEQUENT EXIT OF A WEDGE

In this section, the present numerical method is applied to the high-velocity water entry and subsequent exit of a wedge. The numerical model is similar to the one considered in Section 3.2, but the velocity prescribed to the wedge is now given by eqn (5), following the method used by Breton et al. [23] to experimentally study suction at lower velocities:

$$U = U_{max}\cos(\omega t + \Phi),\qquad(5)$$

where U_{max} is the maximum velocity of the wedge, ω the pulsation, t the time and Φ the phase defined to ensure a variation of U from $-U_{max}$ to U_{max} during the simulation. For the present simulation, the following values of the parameters have been adopted: $U_{max} = 10$ m/s, $\omega = \frac{\pi}{T}$, $T = 15$ ms and $\Phi = \pi$.

The numerical results in terms of hydrodynamic force and pressure are presented in Fig. 7(a) and 7(b), respectively. The hydrodynamic force increases until a maximum value is reached (positive force peak). Then, the force decreases and becomes negative (suction force) as the wedge decelerates. The force reaches a minimum value at the transition between the entry and exit stages, i.e. when the penetration depth reaches its maximum and the velocity of the wedge is null. After, the force gradually returns to 0. The amplitude of the negative force peak is more than 3 times higher than the positive force peak.

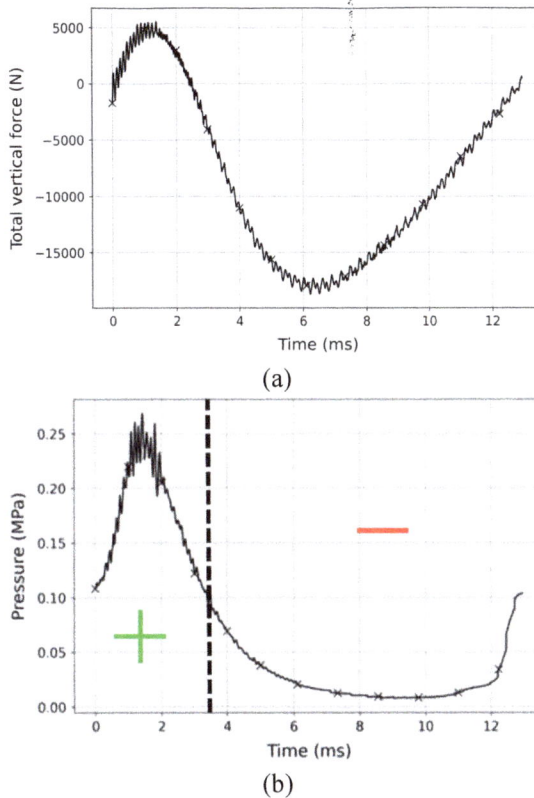

(a)

(b)

Figure 7:   Numerical entry and subsequent exit of the wedge with $U_{max} = 10$ m/s. (a) Time history of the hydrodynamic force; (b) Time-history of the pressure. The probe location is illustrated in Fig. 8. The dashed line separates the stages where the pressure is positive and negative, relatively to $P = 0.1$ MPA.

A pressure peak is measured at the first instants of the simulation, when the water reaches the probe. The probe location is illustrated in Fig. 8. Then, as the penetration depth of the wedge increases and its velocity decreases, the pressure decreases. Around $t = 3.5$ ms, the pressure becomes lower than the ambient pressure ($P_0 = 0.1$ MPa), indicating the occurrence of suction forces, and stabilises around $P \approx 0.008$ MPa. From $t = 10$ ms, the pressure gradually increases to $P = 0.1$ MPA (the ambient pressure). The increase of the pressure corresponds to the opening of the cavity formed under the wedge (ventilation) during the impact (Fig. 8).

The numerical model is able to simulate a negative force and (relative) pressure experienced by the structure at the transition between the entry and exit stages. As the velocity is high, the wedge also experiences a detachment of the fluid (water) and a cavity is formed under the wedge.

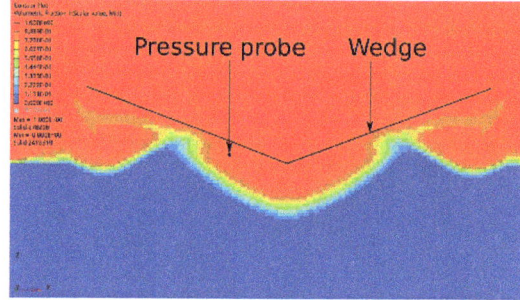

Figure 8: Numerical entry and subsequent exit of the wedge with $U_{max} = 10$ m/s. The figure shows the air volume fraction at $t = 11$ ms. The opening of a cavity under the wedge is visible (ventilation).

5 CONCLUSIONS AND DISCUSSION

In the present article, the numerical simulation of the water entry and water entry and subsequent exit of a wedge have been analysed. The hydrodynamic force and pressure evolutions are emphasised. The study has been carried out numerically using an explicit FE solver (*Radioss*) and a CEL approach.

The numerical results have been compared with existing experimental results for the water entry case. A fairly good agreement has been observed. This study assessed the influence of the most influential numerical parameters on the simulation results. First, varying the contact stiffness has little influence on the numerical hydrodynamic force and pressure evolutions. Second, for the considered element sizes, smaller elements lead to more precise numerical results, but to much higher computation times. The element size also influences the contact stiffness. Therefore, it is a key parameter when modelling hydrodynamic impacts using the present numerical method.

The model also showed its capability to model negative (suction) force and relative pressure. The higher amplitude of the negative force peak further justifies the interest of studying suction phenomena during hydrodynamic impacts representative of realistic industrial applications. However, the numerical model requires validation against experimental results. Therefore, future work will be dedicated to the simulation of experimental cases involving suction forces.

ACKNOWLEDGEMENT

The authors are grateful to ONERA and IFREMER for co-funding this project.

REFERENCES

[1] von Karman, T., The impact on seaplane floats during landing. Report 321, NACA, Aerodynamical Institute of the Technical High School, Aachen, Oct. 1929. Number: NACA-TN-321.

[2]   Wagner, H., Phenomena associated with impacts and sliding on liquid surfaces. Technical report. Translation of Über StoB und Gleitvorgänge an der Oberfläche von Flüssigkeiten. *Z. Angew. Math. Mech.*, **12**, pp. 193–215, 1932.

[3]   Greenhow, M., Wedge entry into initially calm water. *Applied Ocean Research*, **9**, pp. 214–223, 1987.

[4]   Russo, S., Jalalisendi, M., Falcucci, G. & Porfiri, M., Experimental characterization of oblique and asymmetric water entry. *Experimental Thermal and Fluid Science*, **92**, pp. 141–161, 2018.

[5]   Richard, Y., Etudes numériques et expérimentales des impacts hydrodynamiques primaires et secondaires lors du tossage de sections de carènes, Doctoral thesis, ENSTA Bretagne, 2021.

[6]   Cointe, R. & Armand, J.-L., Hydrodynamic impact analysis of a cylinder. *Journal of Offshore Mechanics and Arctic Engineering*, **109**, pp. 237–243, 1987.

[7]   Campbell, I.M.C. & Weynberg, P., Measurement of parameters affecting slamming. Wolfson Unit for Marine Technology and Industrial Aerodynamics, University of Southampton, Report, vol. no. 440, 1980.

[8]   Baldwin, J.L., Vertical water entry of cones. Technical report, Naval Ordnance Laboratory, White Oak, MD, Feb. 1971.

[9]   Mai, T., Greaves, D. & Raby, A., Aeration effects on impact: drop test of a flat plate. *The 24th International Ocean and Polar Engineering Conference*, Busan, Korea, p. 7, 2014.

[10]   Baldwin, J.L. & Steves, H.K., Vertical water entry of spheres. Technical report, Naval Surface Weapons Center, White Oak Laboratory, Silver Spring, MD, May 1975.

[11]   Ribet, H., Laborde, P. & Mahé, M., Numerical modeling of the impact on water of a flexible structure by explicit finite element method: Comparisons with *Radioss* numerical results and experiments. *Aerospace Science and Technology*, **3**, pp. 83–91, 1999.

[12]   Jacques, N., Constantinescu, A., Kerampran, S. & Nême, A., Comparaison de différentes approaches pour la simulation numérique d'impacts hydrodynamiques. *European Journal of Computational Mechanics*, **19**, pp. 743–770, 2010.

[13]   Anghileri, M., Castelletti, L.-M.L., Francesconi, E., Milanese, A. & Pittofrati, M., Survey of numerical approaches to analyse the behavior of a composite skin panel during a water impact. *International Journal of Impact Engineering*, **63**, pp. 43–51, 2014.

[14]   Piro, D.J. & Maki, K.J., Hydroelastic analysis of bodies that enter and exit water. *Journal of Fluids and Structures*, **37**, pp. 134–150, 2013.

[15]   Pentecôte, N. & Vigliotti, A., Crashworthiness of helicopters on water: Test and simulation of a full-scale WG30 impacting on water. *International Journal of Crashworthiness*, **8**, pp. 559–572, 2003. DOI: 10.1533/ijcr.2003.0259.

[16]   Siemann, M.H. & Langrand, B., Coupled fluid-structure computational methods for aircraft ditching simulations: Comparison of ALE-FE and SPH-FE approaches. *Computers and Structures*, **188**, pp. 95–108, 2017.

[17]   Delsart, D., Langrand, B. & Vagnot, A., Evaluation of a Euler/Lagrange coupling method for the ditching simulation of helicopter structures. *5th International Conference on Fluid Structure Interaction*, **105**, Royal Mare Village, Crete, Greece, May, 2009.

[18]   Ortiz, R., Portemont, G., Charles, J. & Sobry, J., Assessment of explicit FE capabilities for full scale coupled fluid/structure aircraft ditching simulations. *23rd International Congress of the Aeronautical Sciences*, Jan., 2002.

[19]   Souli, M. & Sigrist, J.-F., *Interaction Fluide–Structure: Modélisation et Simulation Numérique*, vol. 19, Hermès – Lavoisier, 2010.

[20]   Casadei, F., Leconte, N. & Larcher, M., Strong and weak forms of a fully non-conforming FSI algorithm in fast transient dynamics for blast loading of structures. *EU Science Hub – European Commission*, Corfu, Greece, Nov., p. 20, 2011.

[21]   Bayliss, A. & Turkel, E., Out flow boundary conditions for fluid dynamics. *SIAM Journal on Scientific and Statistical Computing*, **3**(2), 1982. DOI: 10.1137/0903016.

[22]   El Malki Alaoui, A., Nême, A., Tassin, A. & Jacques, N., Experimental study of coefficients during vertical water entry of axisymmetric rigid shapes at constant speeds. *Applied Ocean Research*, **37**, pp. 183–197, 2012.

[23]   Breton, T., Tassin, A. & Jacques, N., Experimental investigation of the water entry and/or exit of axisymmetric bodies. *Journal of Fluid Mechanics*, **901**, A37, 2020. DOI: 10.1017/jfm.2020.559.

ENGINEERING APPROACH TO CALIBRATE A CONCRETE MODEL FOR HIGH SPEED IMPACT APPLICATIONS

HAKIM ABDULHAMID, PAUL DECONINCK & JÉRÔME MESPOULET
Thiot Ingénierie, France

ABSTRACT

This paper describes a study on the mechanical response of concrete under high-velocity impact. It encompasses both experiments and numerical simulations. The aim is to validate an approach for building a concrete numerical model sufficiently robust and accessible to be used for designing civil or defense infrastructures. A conventional concrete (35 MPa compressive strength) has been chosen to apply the method. Experimental tests are conducted to characterize the material in compression and to measure its residual strength during compaction. Impact tests of a kinetic energy projectile (KEP) with an ogive shape nose are also conducted at velocities ranging from 200 to 900 m/s to reproduce both subsonic and supersonic impact conditions. The effect of the concrete confinement is investigated by varying the thickness of a metal jacket surrounding the impacted specimen. Regarding the numerical model, a Holmquist–Johnson–Cook (HJC) for concrete has been calibrated from the measured data. Simulations of the impact perforation are conducted with the γ-SPH solver available in IMPETUS AFEA™. The numerical model has been able to reproduce the main damage in the concrete during the projectile penetration. Good correlation in terms of deceleration profile during penetration is obtained with the experiment. Moreover, the model is robust enough to reproduce the effects of the confinement variation in the projectile residual velocity. This methodology could be applied to other types of concrete materials subjected to various loadings such as near-field blast for example.
Keywords: concrete, KEP, impact, HJC, γ-SPH.

1 INTRODUCTION

Concrete is commonly used as building material for its good strength in compression and relatively low price. However, it shows brittle response in traction unless it is reinforced with metallic bars or fibers. Understanding concrete response facing warhead threats is important for both the design of strategic infrastructure protection and the prediction of warhead performances. From both perspectives, it is important to have access to a robust model that is able to reproduce the main phenomena during such terminal ballistic event. This knowledge also gets increasingly important as terrorisms attacks with improvised explosive devices (IEDs) need to be considered

Many studies related to concrete damage under dynamic events are available in the literature. For the defense industry, the main considered threats are blast and penetration. In general, blast involves a structural response of a wall. However, when the explosive is relatively close, it can lead to similar damage to impact for which the response is highly governed by the concrete damage mechanism and post-failure mechanical properties.

For the case of penetration, one such example is the impact of kinetic energy penetrator (KEP) on reinforced concrete structure. Fig. 1 describes three main steps that are encountered during such event. First, a craterization due to the compaction of the concrete is observed upfront of the projectile. Simultaneously, shock waves are propagating throughout the thickness of the wall. Then, the penetration process starts which is governed by the strength of confined damaged concrete. This step is characterized by the propagation of diffused damage and closing of voids due to compaction. Finally, fracture or spall may appear on the back side as the penetrator is reaching the end the wall. Step two may not be as prevalent if the concrete is not confined. Large crack appears sooner and the penetrator faces less

WIT Transactions on The Built Environment, Vol 209, © 2022 WIT Press
www.witpress.com, ISSN 1743-3509 (on-line)
doi:10.2495/HPSU220111

resistance during its penetration. Therefore, the presence of reinforcement improves the response of the concrete under both traction and compression loadings as it withstands some of the loads and improves the level of confinement [2]. From these descriptions, the residual compression strength of damaged concrete is important to consider for any concrete.

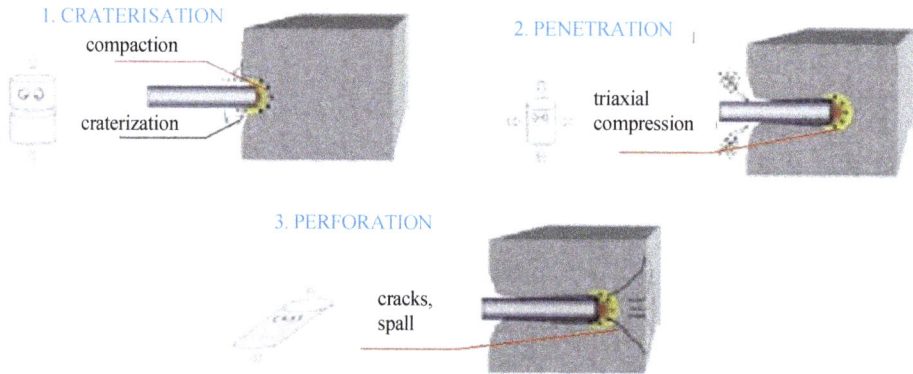

Figure 1: Description of concrete impact phenomena. *(Source: Reproduced from [1].)*

This study focuses in building a concrete model dedicated to impact. The model should be good enough to be used for the defense industry applications and yet it should require a limited number of tests for its calibration. The Holmquist–Johnson–Cook concrete model (HJC) [3] available in IMPETUS AFEA™ corresponds to these criteria. A simplified characterization approach is proposed and tested with a conventional building type concrete. Regarding the discretization method, a meshless method is used to avoid difficulties related to element erosion.

This paper is divided into two parts. Firstly, a description of the modelling approach followed by the presentation of the characterization approach. And finally, the validation test is presented with the corresponding model and an analysis is proposed.

2 MODELLING

2.1 Concrete material model

Apart from the linear response, concrete models for impact simulation must have at least two other characteristics: a brittle response in traction and an important dependency of the compression strength on the confinement for both undamaged and damaged material. Material property can go from a simple elastic-brittle to very complex laws with damage and cracks propagation controls. For example, some of the interesting models for concrete are K&C [4] and RHT (Fig. 2) [5]. They both account for stress triaxiality, strain rates and residual strength when confined. The PRM model [6], [7] and the model proposed by Bazant et al. [8], [9] are even more complex candidates as they consider the propagation of cracks in the damaged material. Such models have been developed for multiple applications (blast, penetration, etc.) and therefore require a quite substantial efforts to be calibrated.

The HJC concrete model [3] has been introduced in 1993 to study the mechanical response of concrete under large deformation, high strain rate and high pressures. The concrete

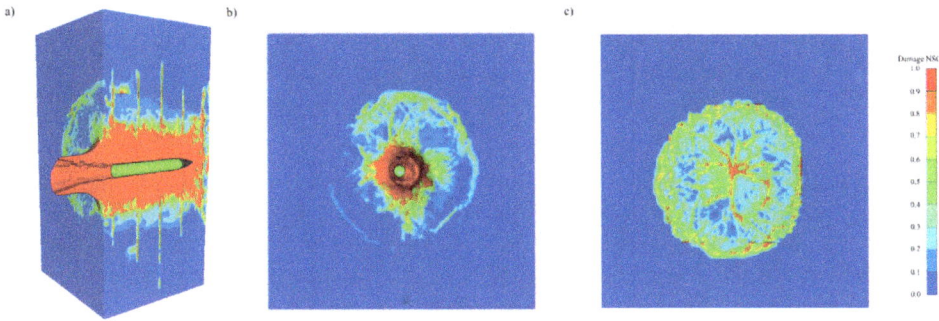

Figure 2: FE modelling damage after 3.9 ms impact of 420 m/s nose projectile on reinforced concrete using RHT model. *(Source: Reproduced from [2].)*

response is described in terms of two independent behaviors: compaction and strength (Fig. 3). The compaction behavior is characterized by three main regions:

- region I: region of linear response with a bulk modulus K_0,
- region II: it is a transition region during which voids are closed,
- region III: fully compacted material where the pressure is computed with K_1, K_2 and K_3.

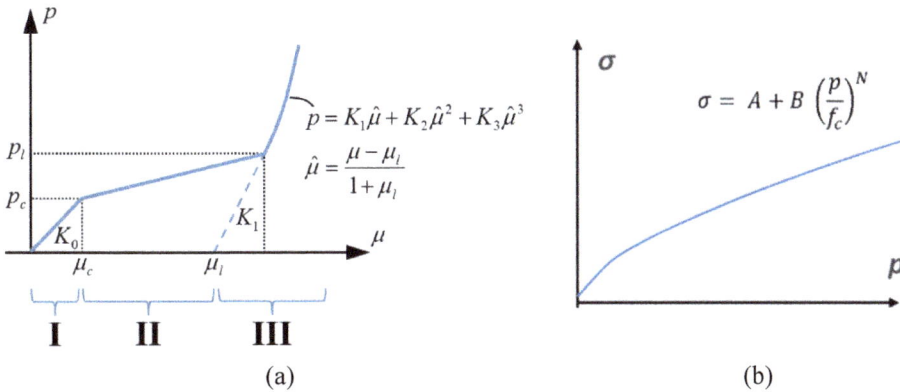

Figure 3: Description of the Holmquist–Johnson–Cook concrete model. (a) Pressure vs volume strain curve; and (b) Yielding curve represented in the equivalent stress vs pressure plane [3].

$$\sigma_y = f_c \left[A \left(1 - D \right) + B \left(\frac{p}{f_c} \right)^N \right] \left[1 + C \ln \left(\frac{\dot{\epsilon}}{\dot{\epsilon_0}} \right) \right]. \tag{1}$$

The material strength is pressure sensitive. It is described by eqn (1) where A, B, N and fc (compressive strength) describe the quasi-static strength of the material. D is a damage variable that progresses during compaction and plastic strain and C is the strain rate sensitivity factor. It can be considered as a simple model in the concrete damage modelling community. Though, it accounts all the major concrete states encountered during a perforation. Another advantage is the availability of the HJC model in major commercial explicit dynamic solvers like LS-DYNA®, IMPETUS AFEATM, Abaqus Explicit®, Ansys Autodyn®.

2.2 Discretization

Most of the presented simulations use Lagrangian finite element with erosion. Some authors have tested meshless methods. For example, Antoniou et al. [10] adapted a discrete element model to simulate confined concrete under hard impact (Fig. 4). The model was able to reproduce some of the cratering damage. SPH is also another attractive meshless method thanks to its ability to handle large arbitrary deformation, though it suffers from instabilities like non-physical oscillation, tensile instabilities, particle clumping.... Questioning their accuracy damage and fragmentation modelling.

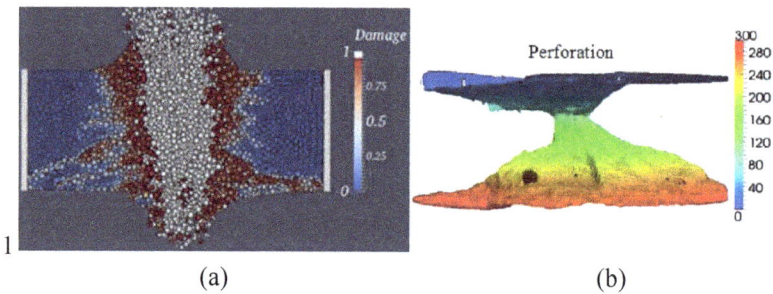

Figure 4:   (a) DEM comparison of perforation crack pattern [10]; and (b) FE modelling damage after 3.9 ms impact of 420 m/s nose projectile on reinforced concrete using RHT model [2].

To achieve robust and consistent stabilization, a new meshless method called γ-SPH based on Arbitrary Lagrangian Eulerian (ALE) considerations is used. The idea is to combine and benefit from both Eulerian and Lagrangian movement descriptions. The formulation has been proven to be conservative, robust, stable and consistent. We refer to Collé et al. [11] for the detailed proof in the context of monophasic barotropic Euler equations (Fig. 5). These properties are validated on many academic test cases ranging from hydrodynamics to solid dynamics [11], [13]. By producing elastic waves free from spurious oscillations and by correcting the tensile instability, γ-SPH provides results in very good agreement with the analytical, experimental and reference numerical ones. It is not only able to handle material failure, but properly capture the strain localization process as well, a precursor to failure. Besides, fracture is now initiated on physical criterion and not numerical instabilities.

2.3 Simplified material characterization approach for the HJC model

As described in the introduction, concrete penetration is mainly characterized by large volume deformation and its high pressure. In the HJC model, the concrete is working in regions I and II. To investigate these areas, two types of characterization tests have been conducted: quasi-static compression tests for the compressive strength (f_c) (Fig. 6) and dynamic oedometric compression to recover the strength vs pressure curve. Both tests have

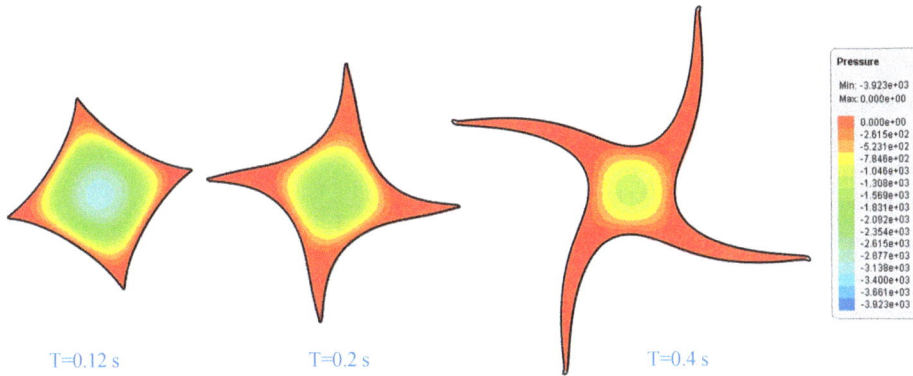

Figure 5:   Simulation of barotropic flows: comparison γ-SPH (in color) and finite element solutions (black border) [12].

Figure 6:   Compression test. (a) JUPITER dynamic press; and (b) Specimen tested in compression.

been conducted with the dynamic press JUPITER (Fig. 6(a)). It has an available volume of 1 m^3 for the specimen which makes it suitable for testing materials with important representative volume element like concrete and geomaterials [14]. The maximum load is 200 kdaN. Depending on the configuration, the press reachable strain rates ranges from 0.1 s^{-1} to 50 s^{-1}. During the test, the displacement of the top plateau is measured with an optoelectronic sensor and the loading force is recorded by a sensor placed on the lower plateau. A preloading of a few kilonewton is applied before the test to catch up any clearances between the plateau and the specimen.

Cylindrical specimens (Ø100 × L200 mm^2) are used for the quasi-static compression test. The measured compression strength is between 35 and 40 MPa. Regarding the oedometric tests, the concrete specimen (Ø75 × L100 mm^2) is confined by a metal jacket as presented in Fig. 7. Apart from the measurement of compression force and displacement, strain gages are used to record the circumferential deformation of the jacket. The analysis of all three curves enables to compute the two main input curves of the model: pressure vs volume strain and the equivalent stress vs pressure (Fig. 8). It should be noted that this last curve is different

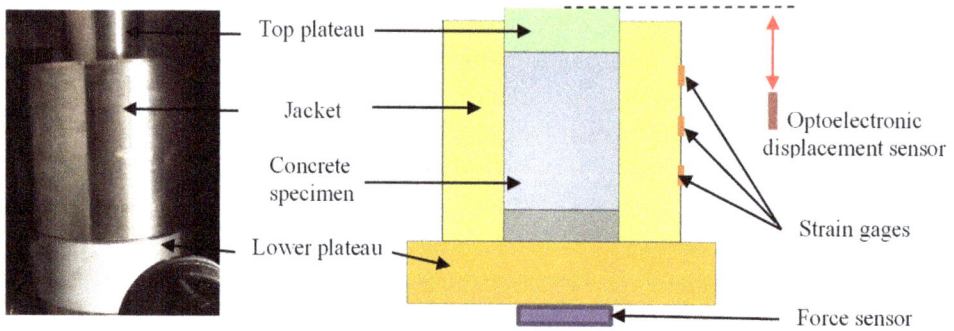

Figure 7: Oedometric test configuration with JUPITER press.

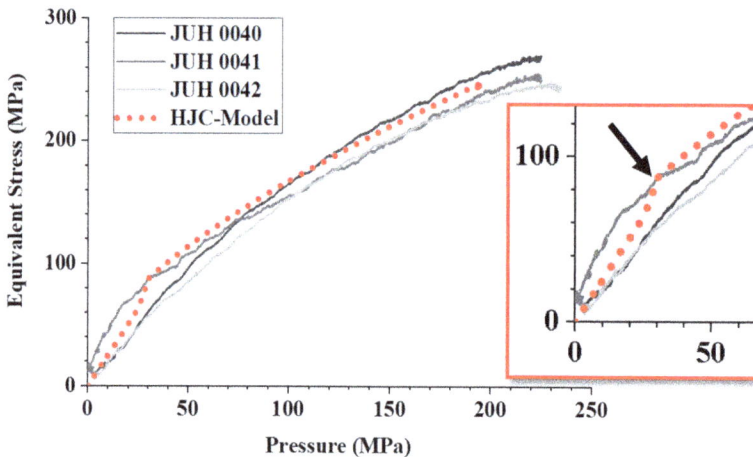

Figure 8:   Equivalent stress vs pressure computed from oedometric tests and the calibrated HJC model curve obtained from the simulation of the test.

from the limit state curve obtained generally through multiple triaxial tests with various loading paths. The equivalent strength obtained from an oedometric test is below the limit state strength. However, from a practical point, it is relatively easier to conduct oedometric tests rather than multiple triaxial tests. Moreover, the oedometric test can be realized at an intermediate or high strain rates regimes.

Experimental curves of equivalent stress vs pressure from three different tests, presented in Fig. 8 show a good repeatability. Then, a fitted HJC model was used to simulate the oedometric compression test above. A simulation of the test is conducted using parameters from the fitted curves. For the simulation, displacement of the lower plateau is constrained and the experimentally measured displacement is applied to the upper plateau. The curve compared with the experimental data in Fig. 8 is obtained from an element in the center of the specimen. The slope discontinuity observed in the beginning of the numerical curve appears at the crushing limit "p_c" of the concrete. This discontinuity is also slightly observed in the experimental curves, though the phenomenon might be smoothed by the inertia of the dynamic press (not reproduced in the model).

3 IMPACT TESTS

3.1 Impact configuration and modelling

The presented impact test can be considered as an intermediate validation test for the model. It helps also to investigate the dynamic interaction between the projectile and the concrete material as this aspect has not been investigated before. It consisted of a down scaled Kinetic Energy Penetrator (KEP) with an ogive nose shape impacting a confined cylindrical concrete specimen (Ø100 mm × 200 mm^2). The concrete is confined by a 5 mm thick metal jacket. To limit the sources of uncertainties, specimens for both characterization and impact tests are made from the same batch of concrete. The KEP is made in high strength 35CrMo15 grade steel and it weighs 520 g. The impact velocity is 396 m/s for the presented test. A high-speed camera records the test from a side view. The test is conducted with a two-stage gas gun in a confinement chamber (Fig. 9).

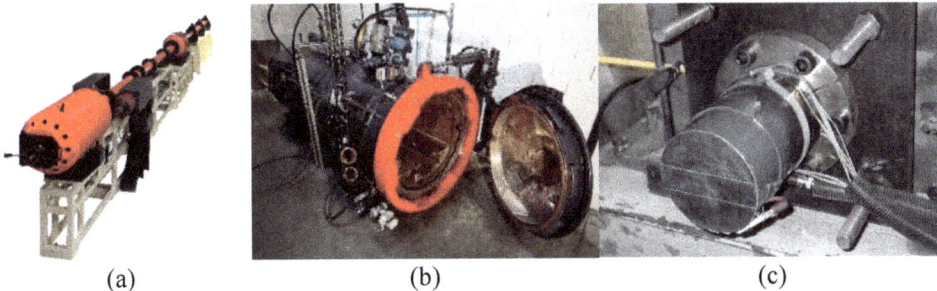

(a)	(b)	(c)

Figure 9:   (a) Two-stage launcher; (b) Containment chamber; and (c) Target specimen configuration inside the chamber.

For the simulation of the impact, the numerical model has integrated all the parts surrounding the concrete specimen as presented in Fig. 10. The concrete is represented with γ-SPH and the remaining parts are modelled with finite elements. A contact interaction is defined between the γ-SPH and the finite elements. There are 1.5 million γ-SPH particles in the model. The KEP is represented with a non-deformable material, its sabot is not geometrically represented in the numerical model. Only the mass of the sabot is considered by modifying the KEP material density.

3.2 Model results analysis

This section begins with a detailed description of the results of a reference case (Fig. 11) conducted at a striking velocity of 396 m/s. As observed in Fig. 11(a), the front side of the specimen is completely damaged. Some concrete has been ejected from the jacket. The specimen is not completely perforated and the KEP is stuck inside. The depth of the penetration of the projectile is 160 mm. On the front side of the target, external diameter of the jacket has increased from 110 to 115 mm.

Regarding the simulation results, the proposed model has reproduced similar projectile penetration (165 mm). The graph in Fig. 11(b) shows the projectile velocity during the

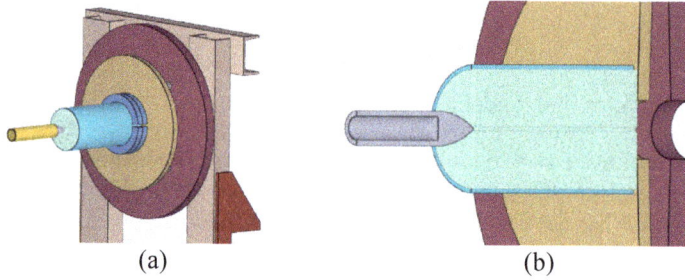

Figure 10:   Isometric view of the numerical model. (a) Large view; and (b) Detail at the KEP and target interaction.

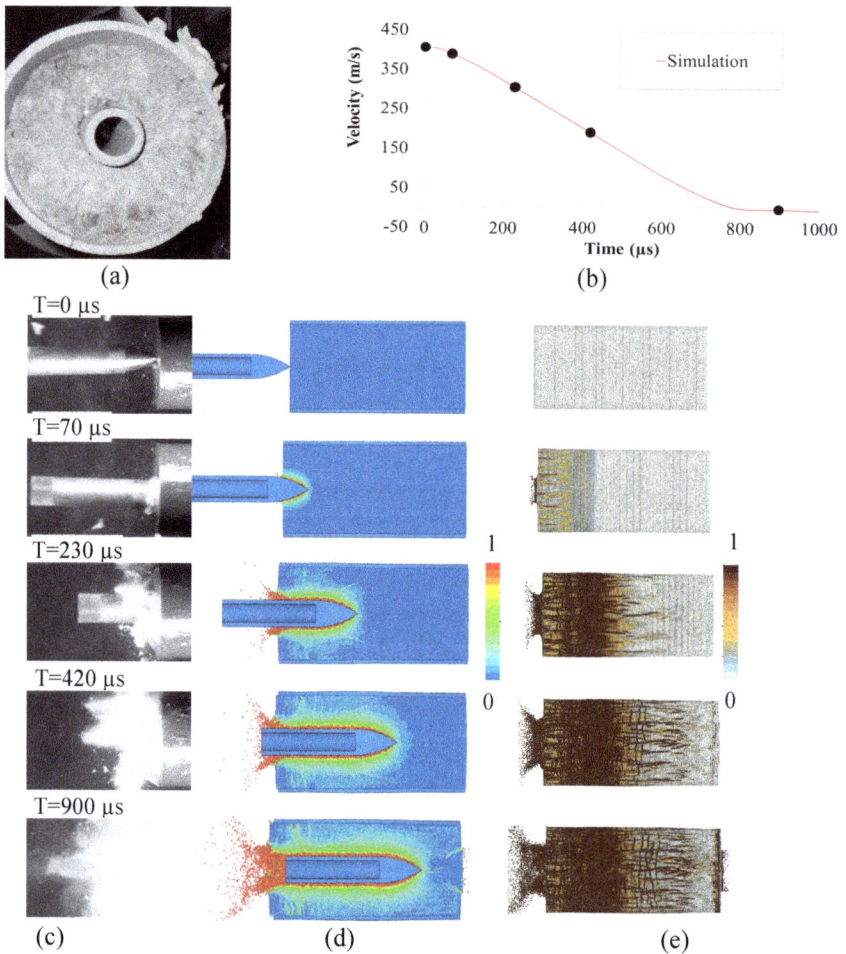

Figure 11:   Comparison of test with simulation. (a) Photography of the impacted specimen; (b) Simulated velocity vs Time profile of the penetrator; (c) Images from high speed camera; (d) Simulated concrete internal strain; and (e) Simulated concrete external damage.

perforation with the points corresponding to the states of the images below. Large deformation is observed around the penetrator due to concrete compaction followed by some radial cracks near the front side of the specimen (Fig. 11(d)). The pressure level of concrete in front of the KEP ranges from 150 to 200 MPa during penetration. On the surface of the concrete specimen, the damage state is characterized by diffuse failure on the front side and fewer large cracks on the right side (Fig. 11(e)). Lack of confinement in the back side inside of the specimen induces shear crack propagation leading to the formation of a spall-like fragment. Some plastic deformation of the jacket is also observed leading to an increase of its diameter of 112.5 mm compared to 115 mm (experimental value).

Many other impact tests have been simulated to evaluate the ability of the model to predict the perforation of the specimen. Fig. 12 compares the measured residual velocity vs striking velocity with the model prediction. Tests have been conducted for striking velocity between 280 and 900 m/s. Experimental critical velocity is between 400 and 550 m/s and the model gives a critical velocity of 490 m/s. The model is also showing very good correlation at higher striking velocity. Another interesting case is the impact of the specimen with thinner jacket (2 mm) for which the simulated residual velocity is coherent with the test. All these highlight the quality of the model and the effectiveness of the approach. The major mechanical phenomena have been considered in the material law and the robustness of the discretization has helped in building such good quality model. As erosion is not needed, the contact between the KEP and the concrete and the concrete confinement are well reproduced all along the penetration process.

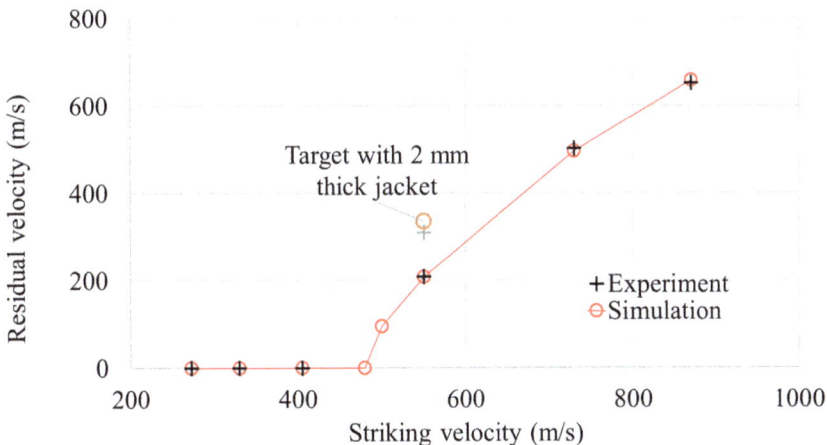

Figure 12: Comparison of experiment and model residual velocity vs striking velocity.

4 CONCLUSION

This paper has presented an approach for the characterization of concrete under impact for building a Holmquist–Johnson–Cook model. HJC is chosen for its relative simplicity and reduced number of parameters. The goal was to develop a relatively robust numerical tool that requires a limited number of tests for it to be applicable in an industrial context. The proposed calibration method uses data from quasi-static compression and oedometric tests to fit the parameters of the material model. Then, an impact configuration test at a reduced scale is proposed to verify and/or improve the response of the model. Furthermore, a robust meshless method (γ-SPH) is used to limits the influence of meshing choice and erosion. The use of such discretization is important because it improves the modelling of projectile/target contact interaction. The concrete confinement is also better reproduced since no element is deleted during the KEP penetration.

First simulations give very good results in terms of penetrator/concrete interaction. Projectile depth of penetration is reproduced by the model as well as the concrete damage. A good balance between model complexity and prediction capability has been reached.

Simulation of impact tests realized at a velocity ranging from 200 to 900 m/s give good correlation in terms of residual velocity. The prediction for a less confined configuration is also coherent with the test. Such model opens the door for simulations of more realistic configuration like concrete with reinforcement. The characterization approach could also be adapted to consider the specific tensile behavior of such concrete materials. The results of such simulation could then be used for improving the performance of KEP or the strength of a wall structure. Information regarding the level of deceleration can be of interest for the design of intelligent systems that include electronics and/or fuzes for KEP filled with explosive. For this particular purpose, it could be very interesting to acquire data from an embedded recording system inside the impactor to monitor its deceleration signature.

ACKNOWLEDGEMENTS
The authors would like to thank Thiot Ingénierie Shock Physics lab team for the realizing all the tests and Thiot Ingénierie for the funding.

REFERENCES
[1] Bailly, P., Tombini, C. & Le Vu, O., Modélisation de géomatériaux sous sollicitations dynamiques élevées: Un tir de pénétration sur cible en béton. *In Colloque du réseau GEO*, Aussois, France, 19, 20, 185, 1996.
[2] Hansson, H., Warhead penetration in concrete protective structures. Thesis, Civil and Architectural Engineering, Stockholm, Sweden, 2011.
[3] Holmquist, T.J., Johnson, G.R. & Cook, W., A computational constitutive model for concrete subjected to large strains, high strain rates and high pressures. *14th International Symposium on Ballistics,* Québec, Canada, 1993.
[4] Malvar, L.J., Crawford, J.E., Wesevich, J.W. & Simons, D., A plasticity concrete material for DYNA3D. *International Journal of Impact Engineering,* **19**, pp. 847–873, 1997.
[5] Riedel, W., 10 years of RHT: A review of concrete modelling and hydrocode applications. *Predictive Modelling of Dynamic Processes*, ed. S. Hiermaier, Springer: London, pp. 143–165, 2009.
[6] Rouquand, A., Présentation d'un modèle de comportement des géomatériaux, applications au calcul de structures et aux effets des armes conventionnelles. Centre d'Etudes de Gramat, technical report, T2005-00021/CEG/NC, 2005.
[7] Pontiroli, C., Rouquand, A. & Mazars, J., Predicting concrete behaviour from quasi-static loading to hypervelocity impact: An overview of PRM Model. *European Journal of Environmental and Civil Engineering*, 2010.
[8] Bazant, Z.P., Xiang, Y. & Prat, P.C., Microplane model for concrete I: Stress–strain boundaries and finite strain. *Journal of Engineering Mechanics,* **122**(3), pp. 245–254, 1996.
[9] Bazant, Z.P., Xiang, Y., Adley, M.D., Prat, P.C. & Akers, S.A., Microplane model for concrete II: Data delocalization and verification. *Journal of Engineering Mechanics,* **122**(3), pp. 255–261, 1996.

[10] Antoniou, A., Daudeville, L., Marin, P., Omar, A. & Potapov, S., Discrete element modelling of concrete structures under hard impact by ogive-nose steel projectiles. *European Physical Journal Special Topics*, **227**, pp. 143–154, 2018.

[11] Collé, A., Limido, J. & Vila, J.-P., An accurate multi-regime SPH scheme for barotropic flows. *Journal of Computational Physics*, **388**, pp. 561–600, 2009.

[12] Collé, A., Limido, J. & Vila, J.-P., An accurate SPH-ALE scheme for barotropic flows. *Proceedings of the 13th International SPHERIC Workshop*, Galway, Ireland, 2018.

[13] Collé, A., Limido, J. & Villa, J.P., An accurate SPH scheme for dynamic fragmentation modelling. *12th International DYMAT Conference, EPJ Web of Conferences*, **183**, 01030, France, 2018.

[14] Mespoulet, J., Plassard, F. & Héreil, P.-L., Strain rate sensitivity of autoclaved aerated concrete from quasi-static regime to shock loading. *EPJ Web of Conferences 31*, 2015.

SECTION 4
PERFORMANCE AND SUSTAINABILITY OF STRUCTURES

FULL-SCALE REINFORCED CONCRETE SLABS WITH EXTERNAL REINFORCED POLYMER: FIELD TEST AND NUMERICAL COMPARISON

RICARDO CASTEDO[1*], ANASTASIO P. SANTOS[1†], CARLOS REIFARTH[1‡],
MARÍA CHIQUITO[1§], LINA MARIA LÓPEZ[1**], ALEJANDRO PÉREZ-CALDENTEY[2,3††],
SANTIAGO MARTÍNEZ-ALMAJANO[4] & ALEJANDRO ALAÑÓN[5‡‡]
[1]Departmento de Ingeniería Geológica y Minera, Universidad Politécnica de Madrid, Spain
[2]Departmento de Mecánica de Medios Continuos y Teoría de Estructuras,
Universidad Politécnica de Madrid, Spain
[3]FHECOR Consulting Engineers, Spain
[4]Escuela Politécnica Superior del Ejército, Spain
[5]Escuela Politécnica Superior de Ávila, Universidad de Salamanca, Spain

ABSTRACT

Numerical simulation of reinforced concrete (RC) slabs with the addition of an external reinforced polymer (FRP) have been developed and compared with full scale real tests. The size of the slabs was 4.4 x 1.46 m, with a span of 4 m, and a thickness of 15 cm. The slabs were built using concrete of class C25/30, and B500C reinforcing steel. Seven tests were conducted, one at a scaled distance of 0.83 $m/kg^{1/3}$, three at a scaled distance of 0.42 $m/kg^{1/3}$, and three at 0.21 $m/kg^{1/3}$. For the biggest scaled distance, the slab had no extra reinforcement. In the other two cases one of the slabs had no extra reinforcement, while the other two tests were performed with carbon fibre reinforcement (CFRP) and E-glass fibre reinforcement (GFRP) located on the face opposite to the blast. Numerical simulation was performed with LS-DYNA software. The study elements (concrete, steel and reinforcement) have been simulated in a Lagrangian formulation with solid elements, beam elements and shells, respectively. Three concrete models have been used and compared: CSCM, MAT72-R3 and RHT. As for the explosive, the CONWEP-based Load Blast Enhanced (LBE) card was used. Reinforcement with CFRP resulted in a generally reduced damage area on both surfaces. All models show a good correlation with the test results and a non-destructive damage estimation technique when comparing them in terms of damage area.
Keywords: LS-DYNA, blast tests, non-destructive technique, CSCM, RHT, LBE.

1 INTRODUCTION

Terrorist acts have unfortunately increased worldwide over the last century, but especially in the west and against civilian populations the use of improvised explosive devices (IEDs) has multiplied with a total of 11,971 attacks recorded between the years 2010 and 2020 alone [1]. Many of these attacks caused damage to structures (pillars, slabs, or beams) or their enclosures or cladding (curtain walls, bricks, tiles, etc.). The main problems derived are two, the first one a possible collapse of the structure and the consequent increase of victims; the second problem are the projections of materials that can cause serious injuries and even death. To better understand their performance and therefore improve the way to protect them,

[*] http://orcid.org/0000-0002-5725-6272
[†] http://orcid.org/0000-0003-2708-9378
[‡] https://orcid.org/0000-0003-2677-5421
[§] http://orcid.org/0000-0003-2207-6674
[**] http://orcid.org/0000-0001-6250-0893
[††] https://orcid.org/0000-0002-8575-1860
[‡‡] https://orcid.org/0000-0003-3509-5831

WIT Transactions on The Built Environment, Vol 209, © 2022 WIT Press
www.witpress.com, ISSN 1743-3509 (on-line)
doi:10.2495/HPSU220121

numerous studies have focused on the field test and numerical simulation of a structural element with different IEDs and various reinforcements (external or internal). The problem of many of these works is that for budget reasons, limitations in the explosive loads, excessive complexity (in simulations or handling the structural elements), tests are carried out on a small scale with the limitations that this entails when generalizing the conclusions (scale factor) [2].

The ideal would be to study the structures in their "real" dimension, but this is a complicated task and, in many cases, unrealistic due to mentioned limitations. There are some publications on this subject, such as the work by Bermejo et al. [3], which analyses the collapse of a two-span, three-story full-scale structure in the face of the sudden demolition of a central column; or the work by Kernicky et al. [4]. A good summary of different progressive collapse of reinforced concrete structures can be found in the recent work of Alshaikh et al. [5]. It is more common to find works on structural elements such as slabs [6], [7], beams-columns [8], [9], or walls [10], [11]. Of these, the vast majority are on a reduced scale, but there are some publications that present results on a real scale. Focusing on the slabs we can highlight among others the work of Castedo et al. [12] that focuses on the simulation of small loads with smooth particle hydrodynamics (SPH) and validates it with real tests of full-scale slabs; or the work of Ruggiero et al. [13] that performs tests with three full-scale slabs and validates the numerical models based on measured pressure and acceleration to then expand the type of models to be performed. The variety of reinforcements to be used for the improvement of structural elements, especially concrete, is very important and increasingly creative, but it is easy to classify between internal reinforcements (added to the matrix) and external reinforcements (outside the concrete). As an example, two recently published state-of-the-art papers cover an important variety of techniques, materials, and application examples [14], [15].

The analysis and comparison of results is usually based on three main aspects: analytical models, numerical models, and non-destructive techniques (NDTs); in addition to the immediate results measured in the tests, such as pressures, accelerations, velocities, or deformations. Analytical models consist of a sequence of formulas that, once adjusted, predict in a more or less simple way some variable of interest such as deformation, deflection, etc. [16]. Numerical models can be of different types depending on how the simulation of the blast load and the structural element is approached. If the load is simulated explicitly (geometry and equation of state), SPH [12] or Arbitrary–Lagrangian–Eulerian (ALE) [17] methods can be used. If the load is simulated using the TNT equivalent, the technique based on numerous blast test (load blast enhanced (LBE)) can be applied [18]. Structural elements can be simulated in the same way with solid elements without considering the surrounding air (Lagrangian or SPH) [19] or taking it into account with ALE [20]. It seems obvious that SPH or ALE techniques involve more computational time and resources but depending on the case, their characteristics are basic for a good and useful simulation. Regarding the NDTs there are a lot of different options depending on the material characteristics, size, or conservation state [21]. However, all these techniques are used to provide reproducible measures of material quality (i.e., concrete) in a specimen (or even a structure) without causing an extra damage to the element or the structure from which it is taken.

In this work, seven real-size RC slabs (in some cases with external reinforcement) against near detonations are presented and compared with numerical simulations using LS-DYNA software version 971-R11 and a non-destructive damage evaluation developed by the working team [22].

2 EXPERIMENTAL TESTS

Seven tests were conducted during 2019 at "La Marañosa" a technological centre of the National Institute of Aerospace Technology (INTA). Some of the most important test parameters are listed in Table 1. It should be noted that the S3 test was performed twice due to minor problems with the measuring equipment during the first trial.

Table 1: Test details.

Test #	Reinforcement	Scaled distance (m/kg$^{1/3}$)	Equivalent mass TNT (kg)	Charge height (m)
S1	None	0.83	1.74	1
S2	None	0.42	13.05	1
S3	Carbon	0.42	13.05	1
S4	Glass	0.42	13.05	1
S5	None	0.21	13.05	0.5
S6	Carbon	0.21	13.05	0.5
S7	Glass	0.21	13.05	0.5

The size of the slabs was 4.40 × 1.46 × 0.15 m and they were placed on concrete supports (2 × 1.2 × 0.90 m long, wide, high) covered with steel plates. The test S1 was designed to test the measuring equipment and to validate the numerical models, for these reasons the explosive mass was low. The shape of the explosive charge, for the high charge tests, could be defined as something between cube, cylindrical and bag following the UFC [23]. The charge was hanged on a rope for the trial at 1 m distance, while it was placed on top of a piece of polystyrene in the tests at 0.5 m. To estimate the TNT equivalent mass of the dynamite used, a peak pressure equivalent of 0.86 was applied [24]. See Fig. 1 for a general view of the slabs and details of the charge used.

Figure 1: (A) Charge at 1 m; and (B) Charge at 0.5 m.

The properties of the concrete used in all the slabs are density (2,300 kg/m^3), uniaxial compressive strength (UCS) (25 MPa) and maximum aggregate size (0.02 m). The steel reinforcement was designed separately from the bottom (opposite face of the blast) and top (face that directly receives the blast). The 12 mm bars at the bottom were spaced 150 mm apart in both directions, while 10 mm bars at the top were spaced twice as far apart as on the other side. The steel used was the usual B-500 S construction steel. The properties have been extracted from the standard RC 1247/2008 – EHE-08 national design code based on EN-1992 [25]. These are the density (7,850 kg/m³), Young's modulus (200 GPa), tangent modulus (20 GPa), yield strength (500 MPa) and Poisson's ratio (0.3).

In one case the RC slab was externally reinforced with carbon fibre (CFRP), in other, the slab was reinforced with E-glass fibres (GFRP). The same methodology has been followed on both scaled distances, first the non-external reinforced slab was tested, then the carbon one and finally the glass one. To ensure the proper functioning of the reinforcements and their solid union with the concrete, the carbon fibre was impregnated with an epoxy resin-based adhesive of thixotropic consistency, while the fibre glass was glued with a polyurethane-based adhesive. The main properties of the fibres are shown in Table 2.

Table 2: Properties of the carbon and glass fibres [19].

Property	Units	Carbon	Glass
Density	kg/m³	1830	2000
Young modulus	GPa	252	42
Shear modulus	GPa	4.8	3.9
Elongation at failure	%	1.8	4
Poisson's ratio	–	0.3	0.285
Compression strength, principal direction	GPa	0.298	0.1129
Tensile strength, principal direction	GPa	2.93	1.62
In-plane shear strength	GPa	0.0845	0.0308

3 NUMERICAL MODEL

The 3D numerical models developed in this work are based on the use of LBE for the explosive loading, while the slab concrete, steel reinforcement and fibre reinforcement in a Lagrangian technique. The concrete has been simulated with solid elements (156936 elements), the steel reinforcement with beam elements (7,024 elements), and the fibres with shells of thickness equal to the real fibres, i.e., 1 mm (16,614 elements). Despite the importance of mesh convergence, 18 × 18 × 18 mm and 50 mm meshes have been used in this work for the solid and beam elements, respectively. An 18 × 18 mm mesh has been used for the shell (fibres), as in the solid elements, to have coincident nodes. These data are based on previous works already published by Castedo et al. [12], Reifarth et al. [19] and Alañón et al. [26]. It should be noted that the model has been made in a complete manner, without taking advantage of any of the symmetries, due to the absence of one of the reinforcing bars (see Fig. 2). This was detected once the tests were done and due to the spalling of the concrete that left the reinforcement visible. See Reifarth et al. [19] for more details.

In this work, three different concrete models have been used: the Continuous Surface Cap Model (MAT_159-CSCM_concrete), the Riedel–Hiermaier–Thoma (MAT_272-RHT), and the Karagozian and Case (MAT72-R3). Three material models are tested due to the influence of the numerical model, and therefore, of the mathematical treatment of the material characteristics. Furthermore, there is no agreement on the use of one or the other model as the optimal solution. CSCM was developed to simulate roadside safety simulations and was created by the U.S. Federal Highway Administration [27]. RHT was designed to simulate the concrete behaviour exposed to blast loads [28], as well as the MAT72-R3 [29]. Despite their disparate origins, all three models use three failure surfaces corresponding with the elastic limit (yield surface), strength (maximum surface) and failure (residual surface). These three material models have been chosen since they have reduced data input and basically work

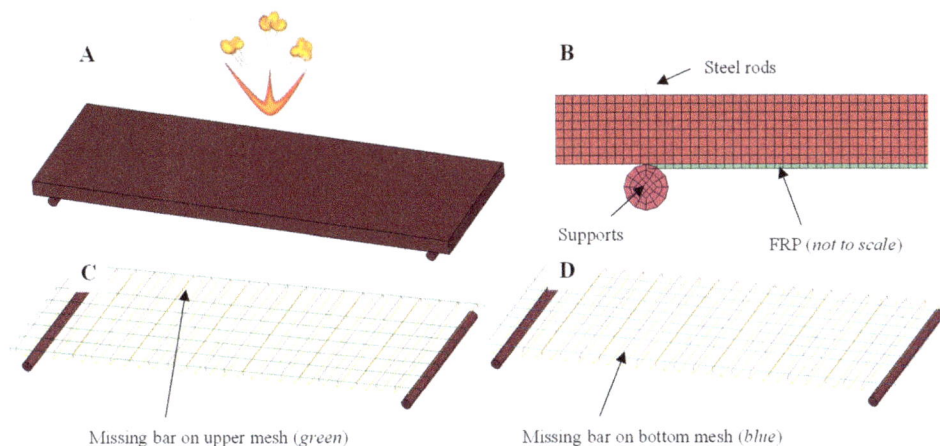

Figure 2: Details of LS-DYNA slab models. (A) Full model; (B) Details of the support and FRP modelling; (C) Longitudinal top and transverse (top = brown and bottom = yellow) steel reinforcement; and (D) Longitudinal bottom and transverse (top = brown and bottom = yellow) steel reinforcement.

with the introduction of UCS, density, and aggregate size. The required parameters, in all cases, are mentioned in Section 2, in the absence of the erosion parameters ERODE (CSCM) and EPSF (RHT). The former is left with its default value (1.05), while for the latter a value of 0.50 comparable to that used by the CSCM is used. Material MAT72-R3 has no damage parameter, although some researchers use the ad-hoc *MAT_ADD_Erosion criterion, it is not the case in this work.

For the incorporation of steel, the piecewise linear plasticity (MAT_24-PLP) model has been chosen. An elasto-plastic behaviour is described with this material model including the strain rate influence on material strength. The stress–strain relation is defined as a bilinear curve controlled by the tangent modulus. The failure is reached by effective plastic strain (7.5%). The parameters are specified in Section 2. One of the characteristics of an event like this is the fast speed at which it occurs, that allows us to reproduce the concrete–steel behaviour as a "solidary" element and incorporate it into the model with the constrained Lagrange in solid tool [30].

The material model called laminated composite fabric (MAT_058) has been used to describe the fibres (carbon (CFRP) and E-glass (GFRP)). Failure criteria is based on Hashin's theory [31]. The failure limit must be defined based on the failure surface selected of the fourth available options, with the smooth failure surface being the final choice after several trials. An erosion parameter equal to –0.55 has been applied after intensive calibration (more details of the FRPs modelling [19]). Material properties are listed in Section 2. The contact with the concrete has been defined using merging nodes (i.e., perfect bond), both elements have the same element size (more details in Reifarth et al. [19]).

In this case the explosive charge has been simulated using its TNT equivalent mass (Table 1) and implemented with the load blast enhanced (LBE) card. Using this card, in addition to the explosive mass, the type of charge, the coordinates of its centre and the face of the slab receiving the explosion must be introduced.

4 NON-DESTRUCTIVE DAMAGE ASSESSMENT (NDDA)

Among the many non-destructive techniques for analysing the condition of materials, here we use a cheap, simple, fast, and effective technique such as the Schmidt hammer. This tool is usually used to estimate the UCS of a material (through the rebound index), in this case concrete, provided that the reinforcement is deep enough (about 5 cm) so as not to affect the results. In this work, the hammer rebound index is used directly, without obtaining the necessary ratios for the estimation of the UCS.

The methodology consists of the creation of small measuring stations with 12 points, six of which are used to measure before the test and six to measure after. The medians of these measurements (Pbt (medians before the test); Pat (medians after the test)) are compared with the Wilcoxon rank-sum test and if the difference is significant at 95%, the measuring station is considered to have had a reduction in the rebound value. The damage (d_H) is calculated as follows for points with significant difference:

$$d_H = 1 - \frac{Pat}{Pbt}.$$

(1)

This parameter can vary between zero (for when the differences are not significant) and one (for when the differences are significant and the "damage is total"). In each slab there are 19 measuring stations distributed according to Fig. 3. To complete the analysis, a mesh of 200 points is created and a damage map is calculated using a triangle based cubic interpolation in MATLAB. For a better assessment of the damage data, and to avoid the heterogeneity of the measurement points, the mean value of the interpolated data is extracted (d_{200}) and used for comparison. See López et al. [22] for detailed information.

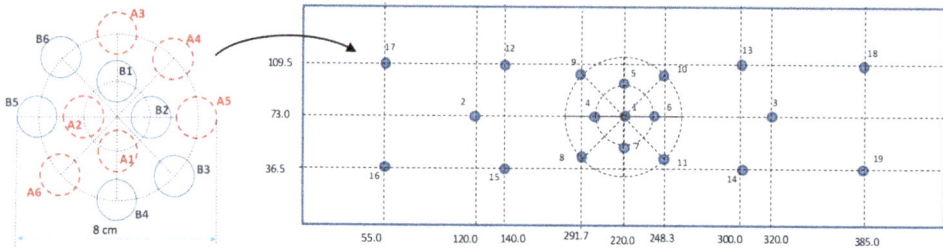

Figure 3: Detail of one of the stations with the six points before (blue) and after (red) the test and location of rebound index measurement stations (19) with the Schmidt hammer. Units in cm.

5 RESULTS AND DISCUSSION

All cases, except the first one, will be analysed according to the scaled distance and based on surface damage area, numerical damage area and non-destructive damage area, see Table 1 for more details. The surface damage area (d_A) is measured by direct test results and can be defined as the eroded area for each surface of the slab divided by the total area of that surface. The numerical damage area (d_{NA}) is measured in the numerical models and calculated as the ratio of the area of damaged elements and the total area for each surface. The non-destructive damage area is extracted from the interpolation maps created in MATLAB and referred as d_{200}, see previous section for details.

5.1 Fitting test (S1).

In this low load test, it was possible to measure the slab acceleration and shock wave pressure without damaging the measuring equipment. This allows us to compare these data with the generated models. The pressure was measured at a radial distance from the load of 1 m, as was done for the acceleration, i.e. 1 m from the centre of the slab. The pressure measured in the field was 2.01 MPa, while the pressure applied with the LBE card is 1.98 MPa, a minimum error of 1.5%. The acceleration measured in the field was equal to 1028.6 g, while the one provided by the CSCM model was 975.53 g (a difference of –5.16%), the RHT was 1,080.10 g (5.01%) and the MAT72-R3 was 1,126.73 (9.54%). Acceleration values are quite good in two of the three cases and acceptable in a third (MAT72-R3). However, as shown in Fig. 4, the damage surfaces of the models are good in CSCM, acceptable in RHT and poor in MAT72-R3. Note that strain rates could only be measured in numerical models and not in tests. For these reasons the MAT72-R3 material is discarded for further analysis. Notice that, in Fig. 4, d_{200} map corresponding to slab S1 is not drawn since the damage obtained in this test was 0%.

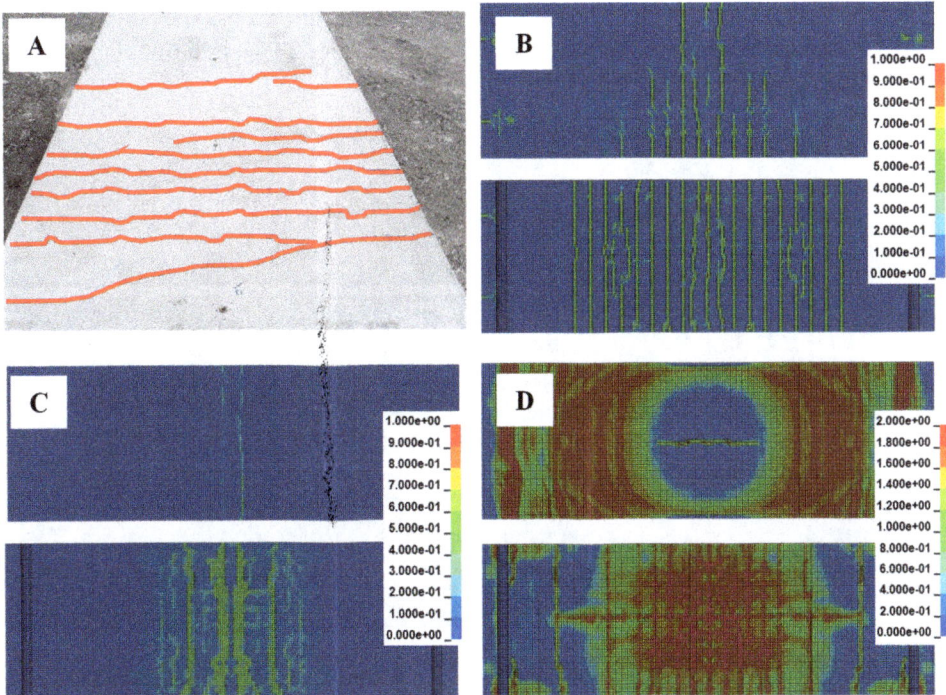

Figure 4: Results for slab S1. (A) Bottom damage; (B) Top and bottom damage with CSCM; (C) Top and bottom damage with RHT; and (D) Top and bottom damage with MAT72-R3.

5.2 Slabs with no external reinforcement (S2, S5).

Considering the results obtained, the simulations of slabs S2 and S5 have been performed with CSCM and RHT. Table 3 shows the damage area values obtained from the tests and the models, in addition to the d_{200} value extracted from the Schmidt hammer maps. The RHT model is not able to reproduce the damage in slab S2 on either side. Given this result, and as slab S5 also shows significant errors especially on the bottom face, the RHT model is discarded for the following simulations with FRPs.

Table 3: Damaged area for slabs S2 and S5 without fibres.

Slab	S	d_A (%)	d_{NA} CSCM (%)	Rel. dif. (%)	d_{NA} RHT (%)	Rel. dif. (%)	d_{200} (%)
S2	T	3.38	4.20	13.80	*	–	18.77
	B	10.34	10.83	4.77	*	–	–
S5	T	8.19	8.48	3.55	9.19	12.17	27.69
	B	18.62	17.63	−5.35	8.82	−52.62	–

Note: S = the surface; T = Top; B = Bottom. * indicates that there is no damage in the model.

As for the damage map generated from NDDA, the S2 slab has a centralized and transversal damage on the surface of the slab that receives the explosion (see also Fig. 5). In slab S5 the damage is 10% higher and is focused on the centre of the slab. Although the damage is presented in an asymmetrical way that may indicate that the load was not perfectly located in the centre or some heterogeneity in it.

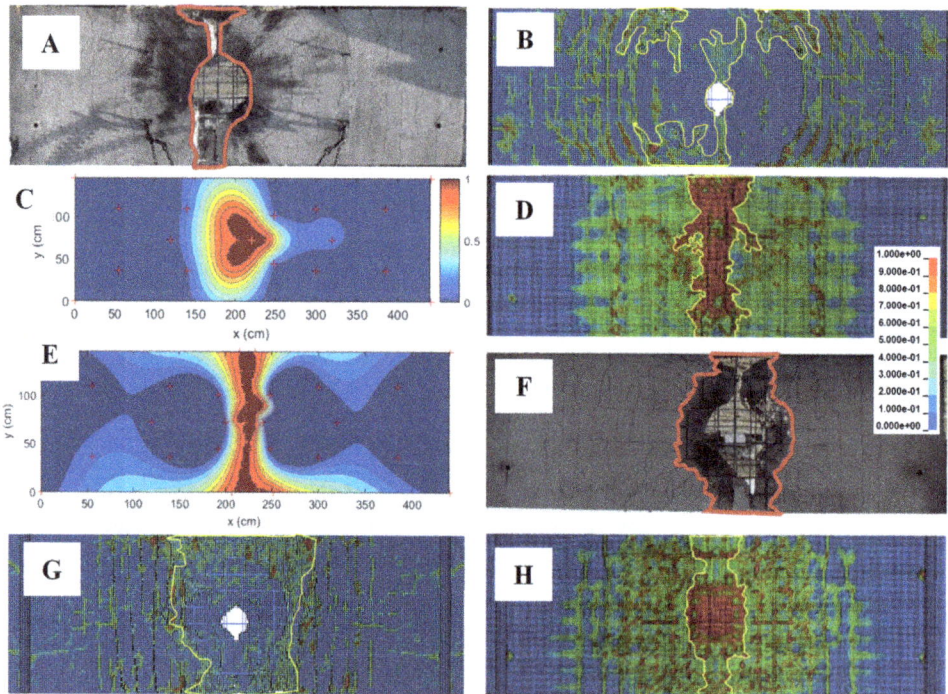

Figure 5: (A) Top damage S5; (B) Top damage with CSCM S5; (C) Damage map from NDDA S5; (D) Top damage with MAT72-R3 S5; (E) Damage map from NDDA S2; (F) Bottom damage S5; (G) Bottom damage with CSCM S5; and (H) Bottom damage with MAT72-R3 S5.

5.3 Scaled distance equal to 0.42 m/kg$^{1/3}$ (S3, S4).

Again, the simulation values are quite good, not exceeding 15% differences with the CSCM model and both types of fibres (see Table 4). In general, the model underestimates the damage area values, except in the case of the slab reinforced with glass fibre. Comparing both results with the S2 slab (without reinforcement), it is observed that the damage on the top face is practically the same, but on the bottom face it is reduced by about 4%–5%.

Table 4: Damaged area for slabs S3 and S4.

Slab	Reinforcement	S	d_A (%)	d_{NA} CSCM (%)	Rel. dif. (%)	d_{200} (%)
S3	Carbon	T	3.27–5.45	3.04	−6.93	19.33–21.26
		B	6.71–6.96	6.43	−7.54	–
S4	Glass	T	3.89	3.66	−6.02	17.85
		B	5.78	6.52	12.72	–

Note: S = the surface; T = Top; B = Bottom.

With respect to the damage maps (d_{200}), the values obtained are much higher than those measured in the test and in the numerical models (see Table 4 and Fig. 6). This is because the data comes from a damage not only superficial, but corresponding to a depth of about 3/5 cm. Therefore, it can detect a loss of resistance that does not seem to be estimated at all with the numerical model. There would be an exception to this, and that is if green areas (with intermediate damage in the model) are considered as damage as is done with the damage maps. In any case, and even if the value of the area were similar, they do not occur in the same areas and with the same pattern.

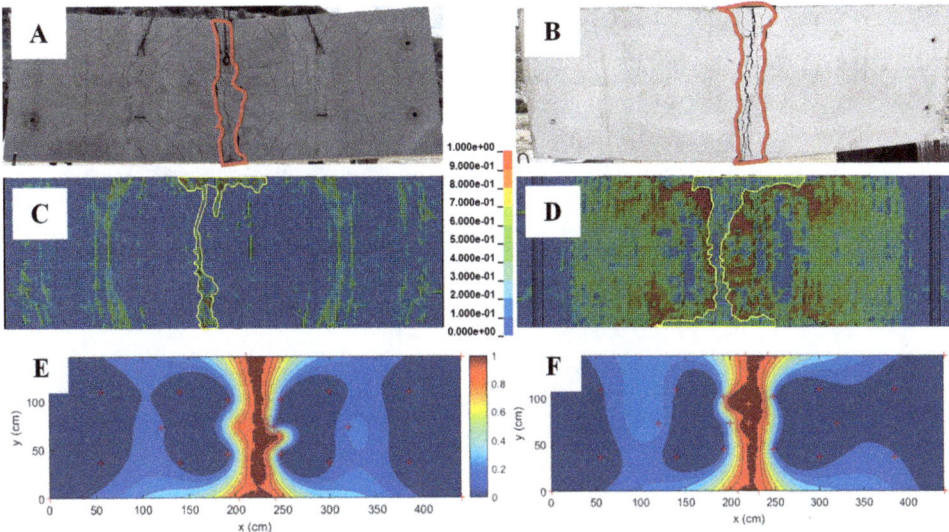

Figure 6:   (A) Top damage slab S4; (B) Bottom damage slab S4; (C) Top damage with CSCM S4; (D) Bottom damage with CSCM S4; (E) Damage map from NDDA S4; and (F) Damage map from NDDA S3.

5.4 Scaled distance equal to 0.21 m/kg$^{1/3}$ (S6, S7).

The damage in the CFRP test (S6) was reduced in comparison with the S5 on both sides (Table 5). However, in the test S7 with the CFRP the top face damage was also reduced but the bottom increased. This is because the glass fibre was not well bonded to the concrete and became detached during the test, dragging with it pieces of concrete that increase the damage surface although not as proper spalling.

Table 5: Damaged area for slabs S6 and S7.

Slab	Reinforcement	S	d_A (%)	d_{NA} CSCM (%)	Rel. dif. (%)	d_{200} (%)
S6	Carbon	T	7.83	7.54	−3.64	28.22
		B	16.78	19.59	16.74	–
S7	Glass	T	7.63	6.96	−8.72	28.93
		B	22.48	22.85	1.66	–

Note: S = the surface; T = Top; B = Bottom.

The numerical model continues to show results very close to reality, with a maximum difference of 15%. The damage maps (d_{200}) continue to have similar values in all three cases (S5, S6 and S7) at the same scaled distance of 0.21 m/kg$^{1/3}$ (Fig. 7). This shows that it can be used to detect the non-visual surface damage, as it is consistent with the calculated damage areas, but they do not allow to differentiate whether a reinforcement improves the constructive element or not (at least on the face receiving the detonation).

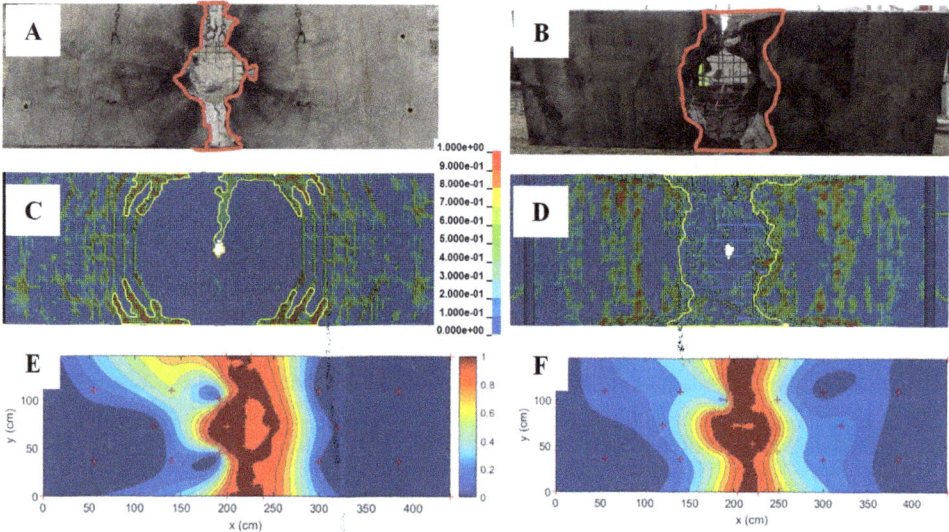

Figure 7: (A) Top damage slab S6; (B) Bottom damage slab S6; (C) Top damage with CSCM S6; (D) Bottom damage with CSCM S6; (E) Damage map from NDDA S6; and (F) Damage map from NDDA S7.

6 CONCLUSIONS

This work allows us to conclude the following:

- reinforcement with carbon or glass fibres improves blast performance especially at intermediate scaled distances, however, a critical point seems to be the bond between the fibres and the concrete;
- damage maps created using Schmidt hammer data respond well to changes in scaled distance, but less so to reinforcement types. However, although more testing is needed, they appear to be a useful tool for quick and reliable analysis at the surface of the material after a detonation;
- 3D numerical models with LS-DYNA software and the appropriate use of material models allow simulations very close to the real results;
- from the above it follows that, given the high cost of real tests, if reliable and calibrated numerical models exist, many more scenarios can be studied at a reduced cost.

ACKNOWLEDGEMENTS

The Centre for Industrial Technological Development (CDTI) funded the PICAEX project, a consortium between TAPUSA, MAPEI and FHECOR, with the collaboration of the Universidad Politécnica de Madrid. The authors would like to thank the staff of "La Marañosa" (INTA) for their support in the trials.

REFERENCES

[1]   Overton, I., Davies, R. & Tumchewics, L., Improvised explosive devices: Past, present and future. *Action on Armed Violence*, 2020.
[2]   Badshah, E., Naseer, A., Ashraf, M., Shah, F. & Akhtar, K., Review of blast loading models, masonry response, and mitigation. *Shock and Vibration*, 2017.
[3]   Bermejo, M., Santos, A.P. & Goicolea, J.M., Development of practical finite element models for collapse of reinforced concrete structures and experimental validation. *Shock and Vibration*, 4636381, 2017.
[4]   Kernicky, T.P., Whelan, M.J., Weggel, D.C. & Rice, C.D., Structural identification and damage characterization of a masonry infill wall in a full-scale building subjected to internal blast load. *Journal of Structural Engineering*, **141**(1), D4014013, 2015.
[5]   Alshaikh, I.M., Bakar, B.A., Alwesabi, E.A. & Akil, H.M., Experimental investigation of the progressive collapse of reinforced concrete structures: An overview. *Structures*, **25**, pp. 881–900, 2020.
[6]   Kumar, V., Kartik, K.V. & Iqbal, M.A., Experimental and numerical investigation of reinforced concrete slabs under blast loading. *Engineering Structures*, **206**, 110125, 2020.
[7]   Foglar, M., Hajek, R., Fladr, J., Pachman, J. & Stoller, J., Full-scale experimental testing of the blast resistance of HPFRC and UHPFRC bridge decks. *Construction and Building Materials*, **145**, pp. 588–601, 2017.
[8]   Liu, Y., Yan, J.B. & Huang, F.L., Behavior of reinforced concrete beams and columns subjected to blast loading. *Defence Technology*, **14**(5), pp. 550–559, 2018.
[9]   Thai, D.K., Nguyen, D.L., Pham, T.H. & Doan, Q.H., Prediction of residual strength of FRC columns under blast loading using the FEM method and regression approach. *Construction and Building Materials*, **276**, 122253, 2021.

[10] Chiquito, M., Castedo, R., Santos, A.P., López, L.M. & Pérez-Caldentey, A., Numerical modelling and experimental validation of the behaviour of brick masonry walls subjected to blast loading. *International Journal of Impact Engineering*, **148**, 103760, 2021.

[11] Chiquito, M., Clubley, S.K., Martinez-Almajano, S., Santos, A.P., Castedo, R. & Lopez, L.M., Numerical and experimental study of unreinforced brick masonry walls subjected to blast loads. *International Journal of Computational Methods and Experimental Measurements*, **9**(4), pp. 296–308, 2021.

[12] Castedo, R., Santos, A.P., Alañón, A., Reifarth, C., Chiquito, M., López, L.M., Martínez-Almajano, S. & Pérez-Caldentey, A., Numerical study and experimental tests on full-scale RC slabs under close-in explosions. *Engineering Structures*, **231**, 111774, 2021.

[13] Ruggiero, A., Bonora, N., Curiale, G., De Muro, S., Iannitti, G., Marfia, S., Sacco, E., Scafati, S. & Testa, G., Full scale experimental tests and numerical model validation of reinforced concrete slab subjected to direct contact explosion. *International Journal of Impact Engineering*, **132**, 103309, 2019.

[14] Draganić, H., Gazić, G. & Varevac, D., Experimental investigation of design and retrofit methods for blast load mitigation: A state-of-the-art review. *Engineering Structures*, **190**, pp. 189–209, 2019.

[15] Goswami, A. & Adhikary, S.D., Retrofitting materials for enhanced blast performance of Structures: Recent advancement and challenges ahead. *Construction and Building Materials*, **204**, pp. 224–243, 2019.

[16] Caldentey, A.P., Diego, Y.G., Fernández, F.A. & Santos, A.P., Testing robustness: A full-scale experimental test on a two-storey reinforced concrete frame with solid slabs. *Engineering Structures*, **240**, 112411, 2021.

[17] Bisyk, S.P., Chepkov, I.B., Vaskivskyy, M.I., Davydovskyi, L.S., Slyvinskyy, O.A. & Aristarkhov, O.M., Methods for modelling Air blast on structures in LS-DYNA. Comparison and analysis. *Weapons and Military Equipment*, **1**, pp. 22–31, 2019.

[18] Kingery, C. & Bulmash, G., Airblast parameters from TNT spherical air burst and hemispherical surface burst. US Army Armament and Development Center, Ballistic Research Laboratory, 1984.

[19] Reifarth, C., Castedo, R., Santos, A.P., Chiquito, M., López, L.M., Pérez-Caldentey, A., Martínez-Almajano, S. & Alañón, A., Numerical and experimental study of externally reinforced RC slabs using FRPs subjected to close-in blast loads. *International Journal of Impact Engineering*, **156**, 103939, 2021.

[20] Xiao, W., Andrae, M. & Gebbeken, N., Experimental and numerical investigations of shock wave attenuation effects using protective barriers made of steel posts. *Journal of Structural Engineering*, **144**(11), 04018204, 2018.

[21] Helal, J., Sofi, M. & Mendis, P., Non-destructive testing of concrete: A review of methods. *Electronic Journal of Structural Engineering*, **14**(1), pp. 97–105, 2015.

[22] López, L.M., Castedo, R., Chiquito, M., Segarra, P., Sanchidrián, J.A., Santos, A.P. & Navarro, J., Post-blast non-destructive damage assessment on full-scale structural elements. *Journal of Nondestructive Evaluation*, **38**(1), pp. 1–15, 2019.

[23] UFC 3-340-02, Structures to resist the effects of accidental explosions. US Department of the Army, Navy and Air Force Technical Manual, 2008.

[24] Chiquito, M., Castedo, R., López, L.M., Santos, A.P., Mancilla, J.M. & Yenes, J.I., Blast wave characteristics and TNT equivalent of improvised explosive device at small scaled distances. *Defence Science Journal*, **69**(4), pp. 328–335, 2019.

[25] UNE-EN 1992-1-1:2013: Eurocode 2: Design of concrete structures – Part 1-1: General rules and rules for buildings, 2013.

[26] Alañón, A., Cerro-Prada, E., Vázquez-Gallo, M.J. & Santos, A.P., Mesh size effect on finite-element modeling of blast-loaded reinforced concrete slab. *Engineering with Computers*, **34**(4), pp. 649–658, 2018.

[27] Murray, Y.D., User's manual for LS-DYNA concrete material model 159 (No. FHWA-HRT-05-062). Federal Highway Administration. Office of Research, Development, and Technology, USA, 2007.

[28] Livermore Software Technology Corporation (LSTC). LS-DYNA Keyword User's Manual – R11 2018:3186.

[29] Malvar, L.J., Crawford, J.E., Wesevich, J.W. & Simons, D., A plasticity concrete material model for DYNA3D. *International Journal of Impact Engineering*, **19**(9–10), pp. 847–873, 1997.

[30] Castedo, R., Segarra, P., Alañón, A., Lopez, L.M., Santos, A.P. & Sanchidrián, J.A., Air blast resistance of full-scale slabs with different compositions: Numerical modeling and field validation. *International Journal of Impact Engineering*, **86**, pp. 145–156, 2015.

[31] Hashin, Z., Failure criteria for unidirectional fiber composites. *Journal of Applied Mechanics*, **180**, pp. 329–334, 1980.

GROUP ANALYTIC NETWORK PROCESS FOR THE SUSTAINABILITY ASSESSMENT OF BRIDGES NEAR SHORE

IGNACIO J. NAVARRO[1], JOSÉ V. MARTÍ[2] & VÍCTOR YEPES[2]
[1]Department of Construction Engineering, Universitat Politècnica de València, Spain
[2]Institute of Concrete Science and Technology (ICITECH), Universitat Politècnica de València, Spain

ABSTRACT

Since the Paris Agreement was established, great interest has arisen in evaluating the sustainability performance of our structures along with their life cycles. The remarkable economic expenses, the important environmental impacts associated with the construction sector, and the great social benefits that might be derived from a well-designed infrastructure system have put the design of essential infrastructures in the spotlight of many researchers. One of today's main challenges is the derivation of adequate sustainability indicators that aid designers when deciding on the most sustainable design alternative. The sustainability performance of infrastructures is based on various indicators that are often conflicting given their different nature. Consequently, the obtention of such indicators usually needs to be addressed using multi-criteria decision-making methods. The present communication shows the analytic hierarchy process (ANP) for the sustainability assessment of a concrete bridge exposed to a coastal environment, involving several decision-makers. A set of nine quantitative criteria, covering the economic, environmental, and social dimensions of sustainability, has been considered here.

Keywords: life cycle assessment, sustainability, sustainable design, bridges, analytic network process, multi-criteria decision-making, group.

1 INTRODUCTION

There has been a great deal of concern about assessing infrastructure sustainability since the well-known Sustainable Development Goals (SDGs) were recently established in 2015. Such interest is justified since the construction sector is recognized as a major environmental stressor, also responsible for a vast proportion of the yearly budgetary expenses of almost every nation. However, the development of infrastructures is, at the same time, an essential resource for the social and economic wellbeing of the regions. Therefore, the design of infrastructures that effectively contribute to the development of the sustainable society to which we all aspire is becoming a great challenge for engineers and architects, as they need to seek a careful balance between the economic, environmental, and social consequences that result from the infrastructures they design. Recent research has been conducted on structural optimization considering a sustainable approach for several infrastructures, such as bridges [1], [2], earth-retaining walls [3], wind turbine foundations [4], buildings [5], dams [6], or tunnels [7], among others.

Such balance is, however, not evident, as it involves conflicting criteria of different nature, and it needs to maximize the positive impacts of their designs and minimize at the same time the negative ones. Consequently, to address the problem of sustainable design, a multi-criteria decision-making (MCDM) approach is usually adopted. MCDM methods are generally based on a first determination of the relevance of each criterion based on the decision maker's (DM) knowledge and their overview of the problem to be assessed. Once such weights are determined, different MCDM procedures exist to determine an adequate solution according to the DM's understanding of the problem, such as TOPSIS, VIKOR, ELECTRE, and others.

WIT Transactions on The Built Environment, Vol 209, © 2022 WIT Press
www.witpress.com, ISSN 1743-3509 (on-line)
doi:10.2495/HPSU220131

The popular method for deriving criteria weights based on DM knowledge is the analytic hierarchy process (AHP) [8]. To determine the criteria weights according to AHP, the DM needs to make pairwise comparisons judging the relative relevance of each criterion concerning each of the rest. One of the main drawbacks of such a methodology is that its weights are highly subjective while decisive for the final decision. This implies that the resulting decision might be affected by the so-called non-probabilistic uncertainties associated with the ability of the DM to consistently reflect their vision of the problem while making the pairwise comparisons. In addition, the more complex is the decision problem to be assessed, and the greater the number of criteria involved, the lower the DM's ability to make accurate or even meaningful judgements [9], [10]. This is particularly the case in sustainability-related decision-making problems, where different and conflicting criteria are usually involved [11].

Consequently, research has been conducted during the last decades to effectively capture the DM's vision of the problem and reflect it in a meaningful criteria weighting. Mainly two trends stand out when dealing with such problems. On the one hand, studies have been conducted that integrate fuzzy [10], [13], intuitionistic [14], or even neutrosophic logic [15] in the AHP procedure to transform the abovementioned uncertainties into a source of useful information for the decision-making problem. On the other hand, other studies emphasize reducing the complexity of the problem to increase the DM's consistency. A popular trend to streamline the decision-making problem is reducing the number of pairwise comparisons to be conducted, thus making it easier for the DM to make consistent judgements. It shall be noted that both trends are not exclusive, and studies have been conducted combining both approaches [16].

The analytic network process (ANP) is an extension of the AHP that allows considering the relations between criteria. ANP has arisen as an adequate decision-making procedure to address sustainability-related problems [17], [18], as it adequately captures the complexity of sustainability issues. In addition, ANP can serve as an effective tool to simplify the decision problem. It may lead to less and more understandable comparisons that might be easier to address by the DM if the problem is properly formulated.

The present communication shows how the ANP can lead to such results when used to determine the weights of quantitative criteria sets. Here, nine sustainability-related criteria are used to determine the design alternative of a particular infrastructure that mainly contributes to sustainability. The infrastructure chosen for this study is a concrete bridge near the shore, thus exposed to an aggressive environment that will lead to significant maintenance. The sustainability life-cycle performance of five different alternative designs is analyzed, and the decision on the excellent design is conducted based on ANP integrating three DMs.

2 MATERIALS AND METHODS

2.1 The analytic network process

As exposed above, in an AHP-based decision model, the criteria, subcriteria, and alternatives are hierarchical, i.e., there is a linear, one-directional relation between these levels. The ANP, on the contrary, allows for a much wider definition of the relations between components, which are now structured in the form of a network. The different elements of the model, be they criteria, subcriteria, or alternatives, are grouped into so-called clusters. The ANP allows a bidirectional relation between clusters, meaning that some or all the elements in one cluster can depend on the elements in another cluster and vice versa. In addition, the ANP allows considering cluster elements that depend on elements contained in that cluster. Both types of dependences are called outer and inner dependences, respectively, and both can be one- or bidirectional.

The construction of the model network that properly represents the decision-making problem to be addressed is an essential step in an ANP. The DM needs first to determine the alternatives and criteria involved in the problem and adequately define the clusters and establish the relations that they consider that might exist between the model elements. Those network relations are then presented in the form of the so-called influential supermatrix that includes every element of the network (criteria, subcriteria, and alternatives). Each element m_{ij} of this matrix is filled with 1 or 0 values, 1 meaning that the element i is influenced by element j. It must be highlighted that this matrix is not reciprocal, i.e., element i might be influenced by element j but not necessarily the other way round.

Once the influential supermatrix has been constructed, the DM must determine the influence of every element belonging to each cluster on any other element. For each cluster, attention will only be paid to those matrix components that are not zero. Such influence is obtained using the usual AHP method. For example, consider that elements A and B, both belonging to cluster C1, influence element C (Fig. 1). A simple AHP model will be constructed to determine which of the two has a more significant influence on element C.

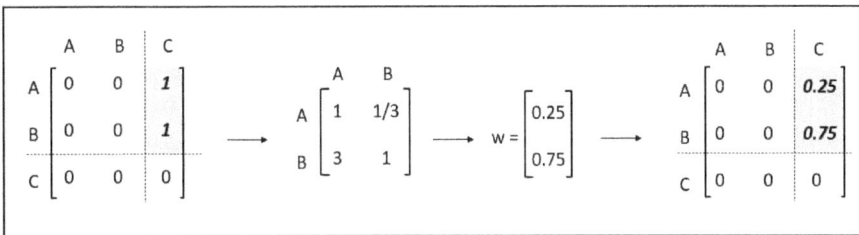

Figure 1: Example of influence determination between elements of a supermatrix.

The DM must fill such a comparison matrix, as usual, using Saaty's fundamental scale to fill a consistent comparison matrix. With every element of the influential supermatrix, a so-called unweighted supermatrix will be constructed. The elements of the influential supermatrix filled with 1 will now be filled with the corresponding weights as shown above (Fig. 1).

It shall be noted that the unweighted supermatrix is not stochastic, i.e., its columns do not sum 1. To make the unweighted supermatrix be stochastic, the elements of each cluster shall be multiplied by the weight of each cluster (considering both criteria and alternatives clusters). These weights are obtained again using a conventional AHP procedure. The resulting stochastic supermatrix is then called the weighted supermatrix.

The last step to determine the criteria weights and the preferred alternatives consists in raising the weighted supermatrix as many times as needed for the elements of each column to converge and remain stable. Such matrix is then called the limiting supermatrix and contains the desired criteria weights and the final rating of the alternatives in each column.

2.2 Group aggregation technique

When several experts are intervening in the decision-making problem, the question arises on including each expert's priorities in the process. Although in recent times complex techniques have been developed to that end, it is common practice to assign each of them a particular voting power and directly aggregate the results obtained by each of them. That voting power is usually determined based on the expert's experience or knowledge in the field [19]. However, as already mentioned above, the more complex a decision problem is, the less accurate and meaningful the expert's judgements, irrespective of their knowledge. The derivation of the experts' voting power proposed here is based on the neutrosophic expert's relevance suggested by [20], [21], where aspects such as the expert's inconsistencies and their manifested self-confidence when emitting judgements are also accounted for.

First, each expert's credibility/knowledge is determined as:

$$\delta_i = \left(\frac{N_i}{\max_{k=1...p}\{N_k\}} + \sum_n K_{c,i}\right)/(n+1), \tag{1}$$

where N_i represents the years of experience of the expert i, and $K_{c,i}$ is a set of n coefficients representing the ith expert's knowledge on the relevant fields to be assessed, and p is the number of experts participating in the decision process. For the sustainability assessment of infrastructures, four coefficients consider their expertise in the social, economic, environmental, and technical assessment of structural designs.

Secondly, the experts' indeterminacy when emitting their judgements is evaluated as:

$$\theta_i = \sum_{q,r=1...n}\left(1 - SC_{qr}^i\right)/M^2, \tag{2}$$

where $SC_{qr}{}^i$ is the average self-confidence expressed by expert i when emitting each pairwise comparison along with the decision-making problem, and M is the total number of judgements emitted.

Lastly, the mean inconsistency of each expert is evaluated based on the inconsistencies derived from each of his/her pairwise comparisons, as:

$$\varepsilon_i = \sum\left(CR_j^i/CR_{lim,j}\right)/J_i, \tag{3}$$

where $CR_i{}^i$ is the consistency ratio of the ith expert regarding the jth comparison matrix filled along the ANP decision process, $CR_{lim,j}$ is the respective limiting consistency ratio which depends on the number of elements to be compared, and J_i is the total number of matrices filled by expert i.

Once these three factors are determined for each expert, the voting power φ_i for an expert i is determined as [22]:

$$\varphi_i = \frac{1-\sqrt{\{(1-\delta_i)^2+\theta_i{}^2+\varepsilon_i{}^2\}/3}}{\sum_{k=1}^{p}\left(1-\sqrt{\{(1-\delta_k)^2+\theta_k{}^2+\varepsilon_k{}^2\}/3}\right)}. \tag{4}$$

Note that if the mean inconsistency of an expert and his/her mean indeterminacy falls to zero, the voting power will be directly proportional to his/her credibility, as usually done in recent research.

3 CASE STUDY

3.1 Description of the functional unit and design alternatives

The methodology described above is used here for the sustainability assessment of five different design alternatives of a concrete bridge deck located in Galicia (Spain) in a coastal environment. The case study presented here is based on Navarro et al. [22]. Besides a conventional baseline design (REF hereafter), three design alternatives are evaluated to prevent chloride-induced corrosion. These reduce the water/cement ratio (alternative W/C35 hereafter), including silica fume or fly ash to the concrete mix, partially substituting the original cement content (alternatives FA20 and HS10, respectively). At last, an alternative with the baseline concrete mix but with galvanized steel reinforcement will also be analysed against its sustainability response along its life cycle (alternative GALV hereafter). Table 1 shows the analysed concrete mixes for each alternative.

Table 1: Concrete mixes for each design alternative.

Concrete mix	REF/GALV	W/C35	SF10	FA20
Cement (kg/m^3)	350	350	280	329
Water (l/m^3)	140	122	140	140
Gravel (kg/m^3)	1,017	1,037	1,017	1,017
Sand (kg/m^3)	1,068	1,095	1,129	1,086
Silica fume (kg/m^3)	–	–	35	–
Fly ash (kg/m^3)	–	–	–	70
Plasticiser (kg/m^3)	5.25	7	4.20	4.94

The functional unit considered here for evaluating the life cycle economic, environmental, and social impacts of each of the abovementioned design alternatives is a 1 m long and 12 m wide bridge deck, including the maintenance operations required to guarantee a service life of 100 years.

In the present life cycle analysis of the abovementioned design alternatives, the maintenance needs for each are different depending on their durability against chlorides. Periodical maintenance is chosen for each of them so that the probability of failure at the year when preventive maintenance takes place is less than 10%. For the present analysis, failure is considered when the chloride content at the rebar depth exceeds the critical chloride threshold. Table 2 presents the parameters assumed for the reliability analysis and the maintenance period chosen for each alternative.

Table 2: Durability parameters for the calculation of each alternative's reliability.

Parameter	REF	GALV	W/C35	SF10	FA20
D_0 ($\times 10^{-12}$ m^2/s)	8.90 (0.90)	8.90 (0.90)	5.80 (0.47)	1.23 (0.17)	4.65 (0.35)
C_{cr} (%)	0.60 (0.10)	1.20 (0.21)	0.60 (0.10)	0.60 (0.03)	0.60 (0.10)
Cover (mm)	40 (2)	40 (2)	40 (2)	40 (2)	40 (2)
Maintenance interval (years)	8	20	15	50	25

Table 2 provides the mean value for each parameter, as well as the standard deviation in brackets.

3.2 Impacts assessment

A set of nine criteria is considered here to quantify the sustainability performance of each alternative, each of them corresponding to one particular type of impact. Table 3 describes the criteria.

Table 3: Decision criteria considered for the sustainability assessment of bridge infrastructures.

Sustainability criterion	Description of the impact	Impact assessment
Construction costs	Economic costs associated to the materials and the construction activities required for the construction of the functional unit	Measured in €. No normalization required
Maintenance costs	Economic costs associated to the materials consumed in maintenance operations	Measured in €. No normalization required. Future costs discounted assuming d = 2%
Damage to human health	Damage to human health derived from the manufacture of the construction materials consumed along the life cycle of the bridge	ReCiPe methodology. Includes increase in respiratory disease, in various cancer types and malnutrition, among others
Damage to ecosystems	Damage to ecosystems and species derived from the manufacture of the construction materials consumed along the life cycle of the bridge	ReCiPe methodology. Includes damage to freshwater species, to terrestrial species and to marine species
Scarcity of natural resources	Consumption of natural resources such as gas or oil derived from the manufacture of the construction materials consumed along the life cycle of the bridge	ReCiPe methodology. Measures the increased extraction costs of oil, gas or coal
Employment generation	Employment generated through the manufacture, construction, and maintenance activities	Indicator based on (Cita social). Takes into account gender issues, fair salary, workers safety and unemployment
Economic wealth generation	Economic inflow to regions where production centres are located	Indicator based on (Cita social). Takes into account the Gross Domestic Product of the regions affected by the product system
Impacts on infrastructure users	Construction and maintenance activities affect the accessibility and the safety of users	Indicator based on (Cita social). Considers the maintenance times and driving speed reduction
Externalities	Noise, dust generation, vibrations and affection to public opinion derived from construction and maintenance activities	Indicator based on (Cita social). Considers maintenance times

The economic, environmental, and social life cycle impacts have been calculated for each alternative considering the same evaluation methodology [22], resulting in the values provided in Table 4. It shall also be noted that the present case study shares the same product system like the one provided in Navarro et al. [22].

4 RESULTS AND DISCUSSION

Following the ANP procedure described in Section 2.1, the decision problem must be converted into a cluster network. Here, four clusters are considered. The first includes the five design alternatives: REF, W/C35, GALV, SC10, and FA20. The second cluster contains the two economic design criteria: construction and maintenance costs. The third cluster includes three environmental criteria: damage to human health, ecosystems, and resource availability. The last cluster contains the four social criteria in Table 3: employment generation, regional wealth increase, affection to users, and negative impacts on public opinion due to externalities derived from maintenance operations.

Table 4: Sustainability assessment results considering all three dimensions of sustainability.

Impact/criterion	REF	GALV	W/C35	SF10	FA20	Units
Construction cost	1296.38	1322.5	2707.73	1566.64	1386.27	€
Maintenance cost	5850.46	2353.45	2121.26	262.67	1624.01	€
Human health	283.92	142.89	151.85	67.17	130.94	Score
Ecosystems	146.93	73.65	75.81	32.14	64.56	Score
Resources	315.2	181.2	190.6	113.3	164.1	Score
Employment	0.681	0.5704	0.5743	0.5096	0.5585	–
Wealth	0.6557	0.4600	0.8007	0.4006	0.4503	–
Users	0.0655	0.1400	0.1568	0.5017	0.1962	–
Externalities	0.0618	0.1363	0.1532	0.4980	0.1959	–

Each DM is then free to establish the outer and inner dependence relations that he/she considers are relevant to the problem. It shall be noted that the DMs start from a pre-established model, where the sustainability of every alternative depends on every criterion, and the value of every criterion depends on every alternative. Fig. 2 shows the network that results from DM 1's view of the problem presented as an influential supermatrix.

	REF	W/C35	GALV	SF10	FA20	C.C.	M.C.	H.H.	Ec.	Res.	Emp.	R.W.	Us.	Ext.
REF	0	0	0	0	0	1	1	1	1	1	1	1	1	1
W/C35	0	0	0	0	0	1	1	1	1	1	1	1	1	1
GALV	0	0	0	0	0	1	1	1	1	1	1	1	1	1
SF10	0	0	0	0	0	1	1	1	1	1	1	1	1	1
FA20	0	0	0	0	0	1	1	1	1	1	1	1	1	1
C.C.	1	1	1	1	1	0	1	0	0	0	0	1	0	0
M.C.	1	1	1	1	1	0	0	0	0	0	0	1	0	0
H.H.	1	1	1	1	1	0	0	0	1	0	0	0	0	0
Ec.	1	1	1	1	1	0	0	1	0	0	0	0	0	0
Res.	1	1	1	1	1	1	1	1	1	0	0	0	0	0
Emp.	1	1	1	1	1	1	1	0	0	0	0	1	0	0
R.W.	1	1	1	1	1	0	0	0	0	0	0	0	0	0
Us.	1	1	1	1	1	0	0	0	0	0	0	0	0	1
Ext.	1	1	1	1	1	0	0	0	0	0	0	0	1	0

Figure 2: Influential supermatrix from DM 1. (C.C. = construction costs; M.C. = maintenance costs; H.H. = human health; Ec. = ecosystems; Res. = resources depletion; Emp. = employment; R.W. = regional wealth; Us. = Users; Ext = externalities.)

Henceforth, and for simplicity, ANP results will be shown only for DM 1. The unweighted supermatrix will be obtained once the influential matrix has been constructed (Fig. 3). It shall be noted that, given that the present problem includes only quantitative criteria, the values of the first five rows and columns of the supermatrix can be obtained straightforwardly from the values presented in Table 4.

In order to obtain a stochastic, weighted supermatrix, the DM is required to determine the weight of the clusters using a conventional AHP procedure. It shall be noted that, in those pairwise comparisons, only those clusters involved are considered, thus simplifying the number of comparisons to be done and therefore increasing consistency (Fig. 4).

	REF	W/C35	GALV	SF10	FA20	C.C.	M.C.	H.H.	Ec.	Res.	Emp.	R.W.	Us.	Ext.
REF	0	0	0	0	0	0.237	0.031	0.089	0.085	0.110	0.235	0.237	0.062	0.059
W/C35	0	0	0	0	0	0.232	0.077	0.177	0.169	0.191	0.197	0.166	0.132	0.130
GALV	0	0	0	0	0	0.113	0.086	0.166	0.165	0.182	0.198	0.289	0.148	0.147
SF10	0	0	0	0	0	0.196	0.693	0.376	0.388	0.306	0.176	0.145	0.473	0.476
FA20	0	0	0	0	0	0.222	0.112	0.193	0.193	0.211	0.193	0.163	0.185	0.187
C.C.	0.819	0.640	0.439	0.144	0.539	0	1	0	0	0	0	0.700	0	0
M.C.	0.181	0.360	0.561	0.856	0.461	0	0	0	0	0	0	0.300	0	0
H.H.	0.261	0.268	0.263	0.272	0.261	0	0	0	0.250	0	0	0	0	0
Ec.	0.504	0.520	0.527	0.567	0.530	0	0	0.700	0	0	0	0	0	0
Res.	0.235	0.211	0.210	0.161	0.209	1	1	0.300	0.750	0	0	0	0	0
Emp.	0.465	0.437	0.341	0.267	0.399	1	1	0	0	0	0	1	0	0
R.W.	0.448	0.352	0.475	0.210	0.321	0	0	0	0	0	0	0	0	0
Us.	0.045	0.107	0.093	0.263	0.140	0	0	0	0	0	0	0	0	1
Ext.	0.042	0.104	0.091	0.261	0.140	0	0	0	0	0	0	0	1	0

Figure 3: Unweighted supermatrix from DM 1.

	REF	W/C35	GALV	SF10	FA20	C.C.	M.C.	H.H.	Ec.	Res.	Emp.	R.W.	Us.	Ext.
REF														
W/C35														
GALV			0				0.4231			0.5729			0.6694	
SF10														
FA20														
C.C.			0.0841				0.1222			0			0.0879	
M.C.														
H.H.														
Ec.			0.7049				0.2274			0.427			0	
Res.														
Emp.														
R.W.			0.2109				0.2274			0			0.2426	
Us.														
Ext.														

Figure 4: Weight of each cluster from DM 1.

	REF	W/C35	GALV	SF10	FA20	C.C.	M.C.	H.H.	Ec.	Res.	Emp.	R.W.	Us.	Ext.
REF	0	0	0	0	0	0.114	0.013	0.051	0.049	0.110	0.235	0.159	0.045	0.043
W/C35	0	0	0	0	0	0.112	0.033	0.101	0.097	0.191	0.197	0.111	0.097	0.096
GALV	0	0	0	0	0	0.055	0.036	0.095	0.094	0.182	0.198	0.194	0.109	0.108
SF10	0	0	0	0	0	0.094	0.293	0.215	0.222	0.306	0.176	0.097	0.347	0.350
FA20	0	0	0	0	0	0.107	0.047	0.110	0.111	0.211	0.193	0.109	0.136	0.138
C.C.	0.069	0.054	0.037	0.012	0.045	0	0.122	0	0	0	0	0.062	0	0
M.C.	0.015	0.030	0.047	0.072	0.039	0	0	0	0	0	0	0.026	0	0
H.H.	0.184	0.189	0.186	0.191	0.184	0	0	0	0.107	0	0	0	0	0
Ec.	0.355	0.367	0.372	0.400	0.374	0	0	0.299	0	0	0	0	0	0
Res.	0.166	0.149	0.148	0.113	0.147	0.259	0.227	0.128	0.320	0	0	0	0	0
Emp.	0.098	0.092	0.072	0.056	0.084	0.259	0.227	0	0	0	0	0.243	0	0
R.W.	0.094	0.074	0.100	0.044	0.068	0	0	0	0	0	0	0	0	0
Us.	0.009	0.023	0.020	0.055	0.030	0	0	0	0	0	0	0	0	0.266
Ext.	0.009	0.022	0.019	0.055	0.029	0	0	0	0	0	0	0	0.266	0
SUM =	1	1	1	1	1	1	1	1	1	1	1	1	1	1

Figure 5: Weighted supermatrix from DM 1.

Considering the clusters weights presented above, the final weighted supermatrix that results from the dependence network developed by DM 1 is shown in Fig. 5.

The final limiting supermatrix is obtained by powering the weighted supermatrix presented above many times as needed, as every column converges to the same values. Fig. 6 shows the limiting supermatrix obtained for DM 1. From this matrix, the weights of each criterion according to DM 1's view of the problem can be derived from rows 6 to 14 once they get normalized.

	REF	W/C35	GALV	SF10	FA20	C.C.	M.C.	H.H.	Ec.	Res.	Emp.	R.W.	Us.	Ext.
REF	0.050	0.050	0.050	0.050	0.050	0.050	0.050	0.050	0.050	0.050	0.050	0.050	0.050	0.050
W/C35	0.074	0.074	0.074	0.074	0.074	0.074	0.074	0.074	0.074	0.074	0.074	0.074	0.074	0.074
GALV	0.074	0.074	0.074	0.074	0.074	0.074	0.074	0.074	0.074	0.074	0.074	0.074	0.074	0.074
SF10	0.138	0.138	0.138	0.138	0.138	0.138	0.138	0.138	0.138	0.138	0.138	0.138	0.138	0.138
FA20	0.082	0.082	0.082	0.082	0.082	0.082	0.082	0.082	0.082	0.082	0.082	0.082	0.082	0.082
C.C.	0.020	0.020	0.020	0.020	0.020	0.020	0.020	0.020	0.020	0.020	0.020	0.020	0.020	0.020
M.C.	0.020	0.020	0.020	0.020	0.020	0.020	0.020	0.020	0.020	0.020	0.020	0.020	0.020	0.020
H.H.	0.098	0.098	0.098	0.098	0.098	0.098	0.098	0.098	0.098	0.098	0.098	0.098	0.098	0.098
Ec.	0.188	0.188	0.188	0.188	0.188	0.188	0.188	0.188	0.188	0.188	0.188	0.188	0.188	0.188
Res.	0.140	0.140	0.140	0.140	0.140	0.140	0.140	0.140	0.140	0.140	0.140	0.140	0.140	0.140
Emp.	0.049	0.049	0.049	0.049	0.049	0.049	0.049	0.049	0.049	0.049	0.049	0.049	0.049	0.049
R.W.	0.029	0.029	0.029	0.029	0.029	0.029	0.029	0.029	0.029	0.029	0.029	0.029	0.029	0.029
Us.	0.019	0.019	0.019	0.019	0.019	0.019	0.019	0.019	0.019	0.019	0.019	0.019	0.019	0.019
Ext.	0.018	0.018	0.018	0.018	0.018	0.018	0.018	0.018	0.018	0.018	0.018	0.018	0.018	0.018

Figure 6:   Limiting supermatrix from DM 1, showing the weights of each criterion as well as the ranking of the alternatives.

On the other hand, the values of the first five rows provide the ranking of the alternatives according to DM 1's judgements. It is observed that the preferred alternative is SF10, namely the one that consists in partially substituting a portion of the cement included in the baseline concrete mix with silica fume.

Lastly, results from each DM shall be aggregated into a final ranking of alternatives. Table 5 provides the characterization of each DM, depending on their knowledge, the self-confidence reported by them while emitting judgements, and the mean consistency when making the pairwise comparisons required by their respective influential supermatrices. The resulting voting power for each of them is also presented.

Table 5:   Characterisation of each DM.

	DM 1	DM 2	DM 3
Years of experience	5	19	15
Knowledge in structural design	0.6	1	1
Knowledge in environmental issues	1	0.4	0.8
Knowledge in economic issues	0.8	0.8	0.4
Knowledge in social issues	0.6	1	0.6
Expert's credibility	0.653	0.840	0.718
Expert's indeterminacy	0.512	0.455	0.424
Expert's inconsistency	0.265	0.270	0.229
Expert's voting power	0.310	0.346	0.344

Considering the above, the scores for each alternative are normalized and aggregated, seeing the relevance of each DM. Table 6 shows the final, aggregated ranking of alternatives.

Table 6: Final scoring of the alternatives.

	REF	W/C35	GALV	SF10	FA20
Aggregated ANP score	0.122	0.176	0.174	0.334	0.195

It is observed that the preferred solution in terms of life cycle sustainability performance is SF10, with a clear advantage if compared to the other design alternative, followed by FA20. Similar results were previously reported by Navarro et al. [22], where design solutions consisting of concrete with silica fume provided the best performances in coastal environments. It is interesting to note the reduced inconsistencies of the DM if compared to the ones reported by Navarro et al. [22]. This is due to the reduced number of comparisons (16 in the case of DM 1, 17 for DM 2, and 18 for DM 3) if compared to the 36 required by a traditional AHP when dealing with a decision problem that includes nine criteria, as the present one.

It shall also be highlighted that ANP allows the DMs to capture their vision of the problem by providing pairwise comparisons and determining the relations they consider relevant to the problem, which can be quite different from one DM to the other.

5 CONCLUSIONS

The construction sector has arisen as an essential tool to reach the sustainable future we all aspire to. It can be responsible for many positive and negative effects on the economy, the environment, and society. However, although crucial to achieving the SDGs recently established, the sustainability assessment of infrastructures still needs further development. The present communication evaluates the sustainability performance of five different bridge deck design alternatives along their life cycle based on the MCDM procedure called ANP. The final ranking of alternatives results from aggregating the judgements of a panel of experts, whose voting power has been determined following a neutrosophic approach.

The preferred design option in terms of its sustainability performance is based on the partial substitution of cement by silica fume. Thus, its durability is increased concerning the conventional baseline design while avoiding part of the negative impacts derived from cement production. Results show the advantages of using ANP when the problem can be formulated based on a quantitative definition of the criteria involved in the decision-making process. In such cases, the ANP methodology reduces the number of judgements to be expressed by the experts and increases their consistency, thus leading to more reliable results.

ACKNOWLEDGEMENT
Grant PID2020-117056RB-I00 funded by MCIN/AEI/ 10.13039/501100011033 and by "ERDF A way of making Europe".

REFERENCES
[1] García-Segura, T., Penadés-Plà, V. & Yepes, V., Sustainable bridge design by metamodel-assisted multi-objective optimization and decision-making under uncertainty. *Journal of Cleaner Production*, **202**, pp. 904–915, 2018. DOI: 10.1016/j.jclepro.2018.08.177.

[2] Navarro, I.J., Yepes, V., Martí, J.V. & González-Vidosa, F., Life cycle impact assessment of corrosion preventive designs applied to prestressed concrete bridge decks. *Journal of Cleaner Production*, **196**, pp. 698–713, 2018. DOI: 10.1016/j.jclepro.2018.06.110.

[3] Molina-Moreno, F., García-Segura, Martí, J.V. & Yepes, V., Optimization of buttressed earth-retaining walls using hybrid harmony search algorithms. *Engineering Structures*, **134**, pp. 205–216, 2017. DOI: 10.1016/j.engstruct.2016.12.042.

[4] Mathern, A., Penadés-Plà, V., Armesto Barros, J. & Yepes, V., Practical metamodel-assisted multi-objective design optimization for improved sustainability and buildability of wind turbine foundations. *Structural and Multidisciplinary Optimization*, **65**, p. 46, 2022. DOI: 10.1007/s00158-021-03154-0.

[5] Sánchez-Garrido, A.J., Navarro, I.J. & Yepes, V., Multi-criteria decision-making applied to the sustainability of building structures based on modern methods of construction. *Journal of Cleaner Production*, **330**, 129724, 2022. DOI: 10.1016/j.jclepro.2021.129724.

[6] Afshar, A., Mariño, M.A., Saadatpour, M. & Afshar, A., Fuzzy TOPSIS multi-criteria decision analysis applied to Karun reservoirs system. *Water Resources Management*, **25**(2), pp. 545–563, 2011. DOI: 10.1007/s11269-010-9713-x.

[7] De la Fuente, A., Blanco, A., Armengou, J. & Aguado, A., Sustainability based-approach to determine the concrete type and reinforcement configuration of TBM tunnels linings. Case study: Extension line to Barcelona Airport T1. *Tunnelling and Underground Space Technology*, **61**, pp. 179–188, 2017. DOI: 10.1016/j.tust.2016.10.008.

[8] Saaty, T.L., *The Analytic Hierarchy Process*, McGraw-Hill: New York, 1980.

[9] Zadeh, L., Outline of a new approach to the analysis of complex systems and decision processes. *IEEE Transactions on Systems, Man., and Cybernetics*, **3**, pp. 28–44, 1973. DOI: 10.1109/TSMC.1973.5408575.

[10] Radwan, N., Senousy, M. & Riad, A., Neutrosophic AHP multi-criteria decision-making method applied on the selection of learning management system. *International Journal of Advancements in Computing Technology*, **8**(5), pp. 95–105, 2016.

[11] Navarro, I.J., Penadés-Plà, V., Martínez-Muñoz, D., Rempling, R. & Yepes, V., Life cycle sustainability assessment for multi-criteria decision making in bridge design: A review. *Journal of Civil Engineering and Management*, **26**(7), pp. 690–704, 2020. DOI: 10.3846/jcem.2020.13599.

[12] Jayawickrama, H.M.M.M., Kulatunga, A.K. & Mathavan, S., Fuzzy AHP based plant sustainability evaluation method. *Procedia Manufacturing*, **8**, pp. 571–578, 2017. DOI: 10.1016/j.promfg.2017.02.073.

[13] Calabrese, A., Costa, R., Levialdi, N. & Menichini, T., Integrating sustainability into strategic decision-making: A fuzzy AHP method for the selection of relevant sustainability issues. *Technological Forecasting and Social Change*, **139**, pp. 155–168, 2019. DOI: 10.1016/j.techfore.2018.11.005.

[14] Ren, P., Xu, Z. & Liao, H., Intuitionistic multiplicative analytic hierarchy process in group decision making. *Computers and Industrial Engineering*, **101**, pp. 513–524, 2016. DOI: 10.1016/j.cie.2016.09.025.

[15] Abdel-Basset, M., Mohamed, M. & Sangaiah, A.K., Neutrosophic AHP-Delphi Group decision making model based on trapezoidal neutrosophic numbers. *Journal of Ambient Intelligence and Humanized Computing*, **9**, pp. 1427–1443, 2018. DOI: 10.1007/s12652-017-0548-7.

[16] Navarro, I.J., Martí, J.V. & Yepes, V., Neutrosophic completion technique for incomplete higher-order AHP comparison matrices. *Mathematics*, **9**(5), p. 496, 2021. DOI: 10.3390/math9050496.

[17] Lam, J.S.L. & Lai, K., Developing environmental sustainability by ANP-QFD approach: The case of shipping operations. *Journal of Cleaner Production*, **105**, pp. 275–284, 2015. DOI: 10.1016/j.jclepro.2014.09.070.

[18] Pourmehdi, M., Paydar, M.M. & Asadi-Gangraj, E., Reaching sustainability through collection center selection considering risk: Using the integration of Fuzzy ANP-TOPSIS and FMEA. *Soft Computing*, **25**, pp. 10885–10899, 2021. DOI: 10.1007/s00500-021-05786-2.

[19] Sierra, L.A., Pellicer, E. & Yepes, V., Method for estimating the social sustainability of infrastructure projects. *Environmental Impact Assessment Review*, **65**, pp. 41–53, 2017. DOI: 10.1016/j.eiar.2017.02.004.

[20] Biswas, P., Pramanik, S. & Giri, B., TOPSIS method for multiattribute group decision-making under single-valued neutrosophic environment. *Neural Computing and Applications*, **27**(3), pp. 727–737, 2016. DOI: 10.1007/s00521-015-1891-2.

[21] Sodenkamp, M.A., Tavana, M. & Di Caprio, D., An aggregation method for solving group multi-criteria decision-making problems with single-valued neutrosophic sets. *Applied Soft Computing*, **71**, pp. 715–727, 2018. DOI: 10.1016/j.asoc.2018.07.020.

[22] Navarro, I.J., Yepes, V. & Martí, J.V., Sustainability assessment of concrete bridge deck designs in coastal environments using neutrosophic criteria weights. *Structure and Infrastructure Engineering*, **16**(7), pp. 949–967, 2020. DOI: 10.1080/15732479.2019.1676791.

INFLUENCE OF THE PROJECTILE SHAPE ON THE DYNAMIC TENSILE CHARACTERIZATION OF CONCRETE USING A SPLIT HOPKINSON BAR

MARIA L. RUIZ-RIPOLL[1], VICTOR REY DE PEDRAZA[2] & CHRISTOPH ROLLER[1]
[1]Fraunhofer Institute for High-Speed Dynamics, Ernst-Mach Institut (EMI), Germany
[2]Departamento de Ciencia de los Materiales, ETSI Caminos, Canales y Puertos,
Universidad Politécnica de Madrid, Spain

ABSTRACT
Because of its relevance in civil infrastructures, the analysis of the dynamic behaviour of concrete has increased exponentially in recent years. This is motivated by the new type of threats that have to be taken into consideration nowadays when designing these types of structures. The growing interest in the dynamic response of concrete arises from the enhancement of its mechanical properties when the material is subjected to high strain rates. In this research, the traditional Split Hopkinson Pressure Bar developed by Kolsky, with a standard compression configuration (including incident and transmitted bars) was modified into a version in which the transmission bar was removed, so that the specimen's response is dominated by tensile stresses inside it. Spalling tests on cylindrical samples were carried out to measure the tensile strength and the fracture energy of conventional concrete. Results for strain rates ranging from 60 to 130 s^{-1} are presented and compared to the respective quasi-static values. As the key point of the research, two different projectile shapes (cylindrical and conical) have also been evaluated, presenting a qualitative and quantitative analysis regarding the variations in tensile stress evolution of the pulses.
Keywords: Split Hopkinson Bar, spalling, dynamic tensile strength, concrete, fracture.

1 INTRODUCTION
The use of the Split Hopkinson Bar (SHB) technique for testing materials subjected to high strain rates has increased notably during the last decades, based on the number of related publications [1]. The interest shown by many researchers in the use of this technique to obtain different material mechanical properties can be appreciated in the variety of test configurations employed. Varying from compression and tensile tests in rocks [2] and concrete [3], compression to tensile testing in metals [4], to most recent 3D configurations to study more complex events [5], this versatility makes the SHB a very powerful tool with many possibilities ahead.

Because of its relevance in civil infrastructures, the analysis of the dynamic behaviour of concrete has increased exponentially during the recent years, researchers are prompted by the new type of threats that nowadays menaces these types of structures. Terrorist attacks, natural disasters, or accidents impulsed researchers to understand the behaviour of concrete under such events. The growing interest in the study of the dynamic response of concrete originates from the enhancement of the mechanical properties when concrete materials are subjected to high strain rates [6] which are still attracting many researchers [7]. For analysing the dynamic behaviour of materials, the traditional Split Hopkinson Pressure Bar device, developed by Kolsky [8] (with the incident and transmitted bars, typically used for studying compressive behaviours) was modified into a version in which the transmission bar was removed, so that the specimen's response is dominated by tensile stresses inside it [9]. This modification opened up a large new field of possibilities on the testing configurations, leading to new several data analysis techniques too. First works came during the 1980s [6], [10], [11], with some important contributions in the 1990s [12], and most recently works

[13]–[16]. However, the complexity of the process of analysis of the results is not an easy task. Variations in the length, geometry, or material of the bars, instrumentation, recording system, and differences in the launch system allow the possibility of a broad range of results and interpretations. The projectile geometry can play an important role on the dynamic characterization of the materials. Based on the work of Gálvez Díaz-Rubio et al. [17] and Rey de Pedraza et al. [18] this work aims to present the influence of two projectile shapes on the determination of the dynamic properties of concrete.

The paper presents the results from a collaboration work between the Polytechnical University of Madrid (UPM) and Fraunhofer EMI. The influence of the projectile geometry (cylindrical and conical shaped) is evaluated for spallation experiments on cylindrical concrete specimens. The experiments are performed in a modified Split Hopkinson Bar at different strain rates. The work presents a quantitative and qualitative analysis concerning the variation in tensile stress evolution of the pulses.

2 EXPERIMENTAL PROGRAM

2.1 Specimen manufacturing

Samples for the dynamic campaign, consisting of concrete cylinders of 75 mm diameter and 300 mm length, were directly obtained by drilling into a uniform concrete block of dimensions $1000 \times 300 \times 250$ mm^3. The drilling of specimens from a large block allows obtaining a very homogeneous mixture for every specimen and avoids the problem of vibration of slender specimens and the wall effect.

The concrete mix was designed by an extern company under the requirements of compressive strength of 50 MPa.

2.2 Quasi-static characterisation

To know the quasi-static behaviour of the concrete mixture, quasi-static compression, Brazilian, and three-point-bending (TPB) tests were conducted. Two specimens per test configuration were performed, the results were averaged and are shown in Table 1. Standard compression and Brazilian tests were performed following the standard specifications [19].

Table 1: Averaged quasi-static concrete properties. Standard deviation (%) in brackets.

Concrete properties	M1	M2	Mean
Density (kg/m^3)	2,288	2,324	2,306 (25)
σ_c (MPa)	65	66.3	65.7 (0.9)
f_t (MPa)	5.2	4.3	4.8 (0.6)
G_F (kN/mm)	3.5E-4	2.5E-4	3.0E-4 (7.0E-5)
$c_{concrete}$ (m/s)	3,800	3,802	3,801 (1)
E_{dyn} (GPa)	33.3	33.3	33.3 (0)

The quasi-static TPB tests were performed as described in RILEM [20], but using cylindrical notched specimens. The total fracture energy (E_F) was then determined using numerical integration of the force-displacement curve, and the area under them. The specific fracture energy (G_F) is derived by dividing the fracture area (A_F), which represents the consumed energy during the fracture process.

2.3 Dynamic test setup

Dynamic spalling tests were carried out, using a modified configuration from the standard compression configuration from a Split Hopkinson Bar [8], (Fig. 1). In the spalling configuration, there is no transmission bar, so the specimen's response is dominated by tensile stresses inside it.

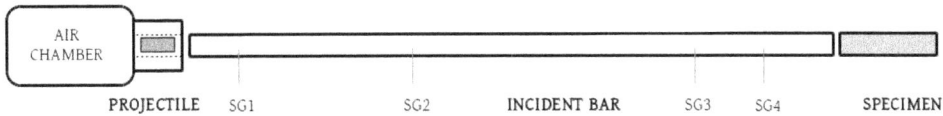

Figure 1: Experimental scheme of the SHB device.

The used SHB device is composed of an aluminium incident bar with 75 mm diameter and 5,500 mm length, and a steel projectile with a length of 60 mm. The projectile (cylindrical or conical shaped) is disposed inside air cannon. The concrete specimen is glued to the incident bar. Strain gauges are displaced along the incident bar and the specimen, tracking the evolution of the wave.

The versatility of the presented configuration has been proved at EMI for several materials. Among others on sedimentary rocks [21], or on high-performance concrete mixes [22].

3 INFLUENCE OF THE PROJECTILE SHAPE

Two different projectile geometries are analysed in this work based on the modification of the classical cylindrical projectile geometry made by Gálvez Díaz-Rubio et al. [17] and Rey de Pedraza et al. [18]. When using cylindrical projectiles with a high length/diameter ratio, the pulse generated can be assumed to be rectangular, as seen in Fig. 2(a). The typical rising part to the constant compression level is modified when the projectile shape changes. Following the work of Kolsky and Shi [23], the use of triangular pulse leads to some advantages after the reflection of the compressive pulse: the stress evolution after reflection follows a gradual increase instead of a sudden rising (Fig. 2(b)), and fewer sections would be simultaneously subjected to equivalent tensile stress.

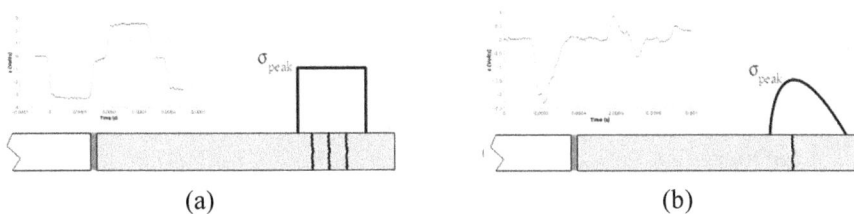

(a)   (b)

Figure 2: Differences in the pulse shapes based on the projectile geometry. (a) Cylindrical projectile; and (b) Conical projectile.

Based on the existing cylindrical projectile, the dimensions of the conical one were defined. Furthermore, before manufacturing it, the behaviour of the two different shapes were validated by using LS-Dyna code. The conical shape projectile was therefore adjusted to the restrains of length and major face diameter of 60 mm (Fig. 3(a)), preserve of impact energy, and the use of polymer sabot (where projectile core is embedded). A comparison between experimental and numerical obtained pulses for both geometries is presented in Fig. 3(b). Conical projectile pulse has a more regular pulse branch in comparison with the disturbed branch of the cylindrical one. By reducing the diameter of the reflecting face of the projectile leads to shorter time-pulses, and therefore to a smaller portion of the sample affected by the superimposition of the waves.

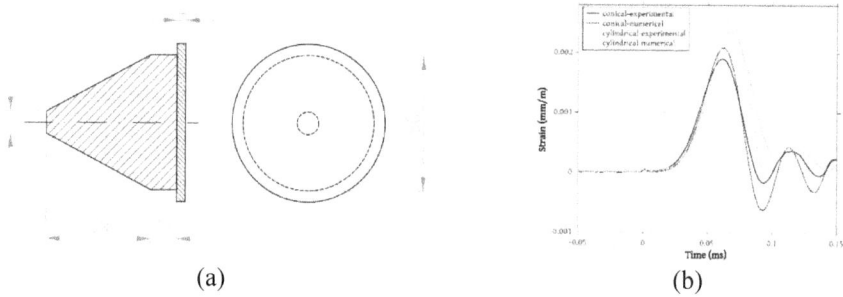

(a)   (b)

Figure 3:   (a) Conic projectile's geometry (dimensions in mm); and (b) Experimental and numerical comparison of the pulse shapes.

4 DATA ANALYSIS APPROACH

The estimation of the tensile strength or the fracture energy required two different configurations of specimens. The difference between them is the use of notched specimens for fracture energy analysis (Fig. 4). Having a notched specimen promotes a single fracture in the specimen, avoiding or reducing the possibility of having multiple fractures as, thanks to the area reduction at the notch, tensile stresses can be kept below the dynamic tensile strength of concrete in every section outside the notch.

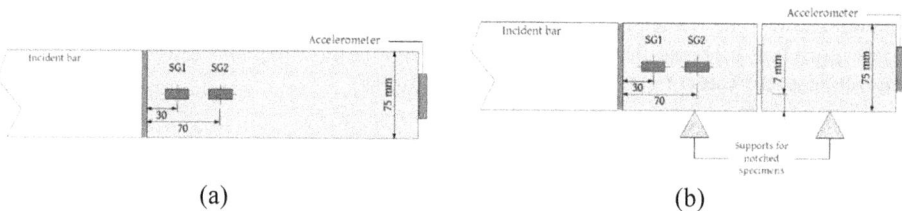

(a)   (b)

Figure 4:   (a) Tensile test specimen configuration; and (b) Energy tests specimen configuration.

In the tensile configuration the specimen was glued to the incident bar using an epoxy resin. For the fracture energy configuration, the specimen is not glued but linked to it using a mortar mix, enough to transmit the compressive pulse from the incident bar to the specimen. The applied methodology (see Section 4.2) requires that all the specimen's parts are free to be ejected after the fracture occurs, so the momentum of each ejected piece of the concrete specimen can be accurately measured which results in a better approximation of the fracture energy. Extra supports are then needed under the specimen, as shown in Fig. 4, to withstand the self-weight of concrete. Two strain gauges are used in both configurations however, the need for two different records for the strains has a special relevance in the case of the energy tests where initial and residual strains inside the concrete specimen have to be recorded at different points to avoid superposition of the pulses.

4.1 Tensile strength

During the spalling test, the projectile is ejected against the incident bar at different air pressures. This pressure influences the stress peak and so the strain rate on the specimen. The tensile strength is derived by using the Novikov approach [24], based on

$$f_{t,dyn} = \frac{1}{2}\rho \cdot c_L \cdot \Delta u_{pb} \tag{1}$$

where the density (ρ) and wave speed (c_L) of concrete and the so-called pullback velocity (Δu_{pb}) of the ejected end of the specimen are used to estimate the tensile strength. The wave speed can be measured thanks to the recorded signals of one of the strain gauges and the accelerometer (see Fig. 4). As the position of both of them is known and the difference in the arrival time is given by the signal records, the wave speed can be directly computed. The pullback velocity can be defined as the difference in velocities of the free end before and after the spallation occurs and is a kinematic approximation of the loss of energy during the fracture process (see Fig. 5). Thus, the dynamic young modulus (E_{dyn}) is derived from the expressions of the theory of unidimensional wave propagation on elastic materials by

$$E_{dyn} = \rho \cdot c_L^2 \tag{2}$$

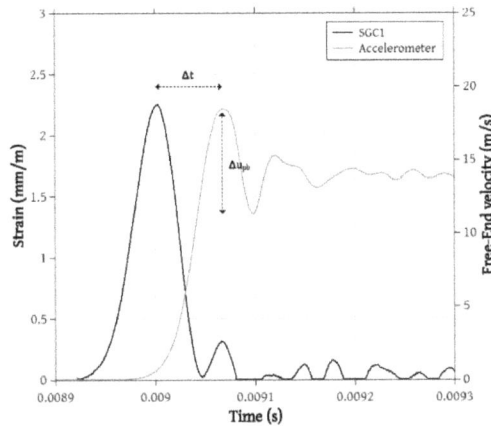

Figure 5: Signal records at the strain gauge and accelerometer.

4.2 Fracture energy

For the analysis of the dynamic fracture energy, the dynamic tensile configuration is modified at some points. First, the use of notched specimens, is required to promote a single and located fracture in the specimen where the area reduction leads to a higher stress level compared to the rest of the sections, ensuring that a fracture is going to occur at that point. Second, the notched specimen is not glued to the incident bar as already commented at the beginning of Section 4.

The dynamic fracture energy was derived by using the Momentum approach, proposed by Schuler et al. [25]. One of the critical factors of this technique is that in notched specimens after the fracture occurs, the formed pieces are ejected freely so that the velocities, and thus the momentum, can be accurately measured. The velocity of the fragments can then be measured by using different sensors, such as digital extensometers, accelerometers or digital image correlation (DIC). As the mass of the separated fragments can be measured after the test, the moment of inertia can be easily computed.

The estimation of the fracture energy (G_F) is derived from different parameters as the momentum transfer between fragments (ΔI), the solid rigid velocity of the ejected free end, before ($v_2(t_1)$) and after fracture occurs ($v_2(t_2)$), and the crack opening velocity ($\dot{\delta}$). The authors suggest to visit Schuler et al. [25] paper for the complete approach details.

5 RESULTS

5.1 Tensile tests

Tensile tests were performed using both projectiles under three different air pressures. For each pressure (0.5, 1.0 and 1.5 bar) two experiments were conducted. Fig. 6 displays the evolution of strain rates as a function of the pressures, reaching strain rates ranging from 60 s^{-1} to 130 s^{-1}. It can be seen how the influence of the projectile shape on the strain rate is minimal, as the strain rates till the spalling moment are very similar for both geometries.

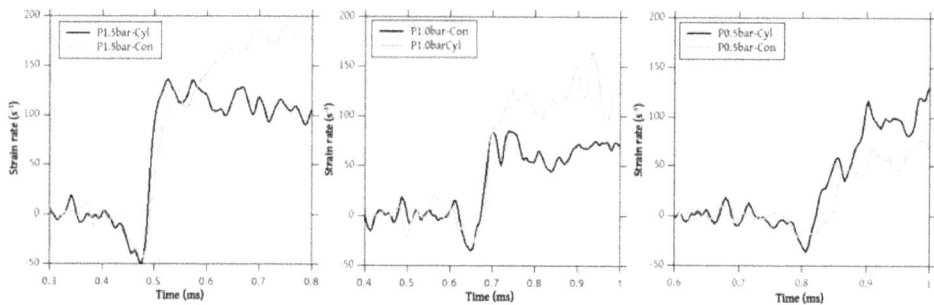

Figure 6:   Evolution of the strain rates during spalling test at different pressures tested (from left to right 1.5, 1.0 and 0.5 bar).

Tensile strenght together with other important parameters are presented in Table 2. The results show uniformity and coherence on the values for all the strain rates analyzed. As expected, the Dynamic Increase Factor (DIF) increase with the strain rate, reaching a value close to 7 at a strain rate of 125 s^{-1}.

An analysis of the fracture pattern was made for a qualitative point of view, comparing the numer and spacing of fractures as a function of the projectile shape and strain rate (see Fig. 7). Specimen tested with the cylindrical projectile presented more damage in comparison to the conical ones. Even when the strain rates where similar for both projectile under same pressure, the specimens tested with the cylindrical projectile displayed a higher density of fractures. The evolution of the facture position, as well as partial opened cracks is more gradual and controlled in the specimens where conical projectile was used. Therefore and based on the fracture patterns, it is convenient the use of conical projectile for spalling test.

Table 2: Dynamic tensile strength results.

#Test	P (bar)	$\dot{\varepsilon}$ (s^{-1})	v $_{min}$(m/s)	v $_{max}$ (m/s)	u $_{pb}$ (m/s)	σ (MPa)	DIF $_{ft}$
Cyl1	1.5	130	11.30	18.46	7.14	31.3	6.5
Cyl2	1.0	88	12.50	18.40	5.86	25.7	5.3
Cyl3	0.5	60	6.24	10.15	3.91	17.1	3.6
Con1	1.5	125	7.00	14.43	7.43	32.6	6.8
Con2	1.0	83	7.20	12.97	5.77	25.3	5.3
Con3	0.5	50	5.18	7.56	2.38	10.4	2.2

(a) (b)

Figure 7: Fracture specimens after spalling tests under different pressures (from upper to lower rows 1.5, 1 and 0.5 bar). (a) Cylindrical projectile; and (b) Conical projectile.

As already comented in previous paragraphs, the gradual progression of the tensile peak after reflection is one of the advantages obtained from using conical projectiles. This gradual growth of stresses lead to a better determiantion of the first crack in time, and it is well differntiated from the rest. This can be seen when comparing the evolution of the strains in Fig. 8(b). Contrarily, in Fig. 8(a), is clear how in the case of cylindrical projectiles, strains arise into a plateau, leading to a wide portion of the specimen being subjected to similar stress at a certain moment in time, causing simultaneous fractures and complicating the identification of the initial crack.

5.2 Fracture energy tests

Notched specimens were tested under pressure below 0.25 bar, three for each projectile shape. Due to the area reduction at the notched section, similar strain rates were 80–100 s^{-1} determined. Results of the dynamic fracture energy parameters are presented in Table 3. As the specimen is designed to be fractured at the notch, only two fragments are generated. For the measurement of the velocities, two time-instants are considered, t_1 corresponding to the

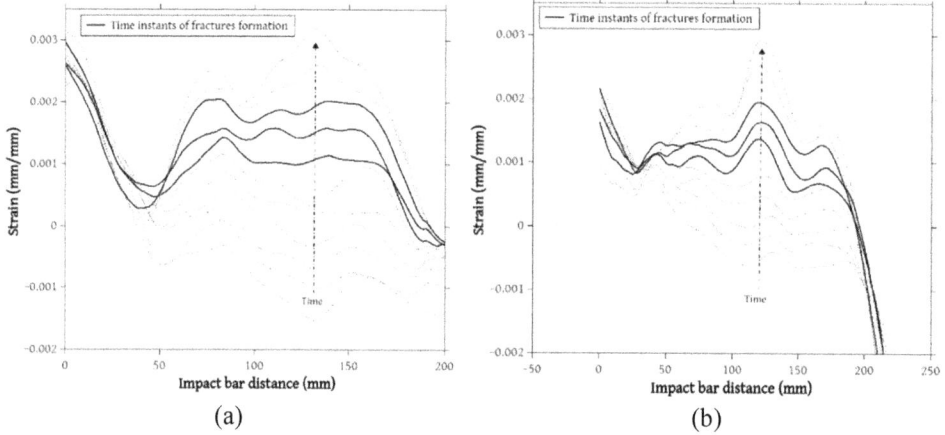

Figure 8: Strain evolution. (a) Cylindrical projectile; and (b) Conical projectile.

moment right before the fracture occurs and t_2 corresponding to the moment just after the fracture. To compute the velocities, in the case of t_1, an analytical procedure was used to derive the velocities of both fragments, while for t_2 velocities were measured using a high-speed extensometer. In the present work, both instants have been measured using a DIC analysis on the images of the high-speed camera. The idea is to obtain a more direct approach to the instant before the fracture.

Table 3: Dynamic fracture energy parameters results.

#Test	$\dot{\varepsilon}$	m_2	ΔI	$\dot{\delta}$	E_F	G_F
	(s^{-1})	(kg)	(kg m/s)	(m/s)	(J)	(N/mm)
Cyl1	86	1.47	1.91	1.52	2.90	1.024
Cyl2	98	1.47	1.62	1.97	3.19	1.127
Cyl3	99	1.48	1.70	2.17	3.69	1.306
Con1	125	1.47	1.87	2.78	5.18	1.832
Con2	85	1.46	1.94	1.46	2.84	1.003
Con3	80	1.50	1.86	1.97	3.65	1.293

Test results show consistency for all performed tested. In test Con1, an alternative configuration to validate the influence of the boundary condition was used. In this case, the specimen was glued to the input bar, ejecting therefore only one part of the specimen after the fracture. An increase in the measured fracture energy for this test can be noted, related either to an inaccurate measurement of the momentum or the increased strain rate for this configuration, but a deeper analysis should be done to clarify this point. From the tests, a mean value of 1.264 N/mm was obtained. Similar values of G_F were obtained between cylindrical and conical projectiles, just a small difference in the higher strain rate produced by the cylindrical projectile. DIF parameter for the fracture energy show that even when the strain rate is kept within a narrow range, a common increasing trend can be seen.

5.3 Dynamic increase factor analysis

As a final point of this research, previous experimental studies are compared with the obtained results. It is important to know wether they fit with literature ones, which have been taken as a reference to check the validity of the results presented.

The dynamic increase factors (DIF) are factors of significant importance in the dynamic analysis of materials and structures, calculated as the ratio of the dynamic to static material parameter, and it is normally defined as the function of strain rate. Fig. 9 presents the DIFs values for different parameters. First, in Fig. 9(a), several tensile strength data for concrete mixtures are compared. The obtainde results in this research are in accordance with the literature data following the well-known trend of linearity till strain rates close to 100 s^{-1}, where the exponential branch with higher values of DIF's starts.

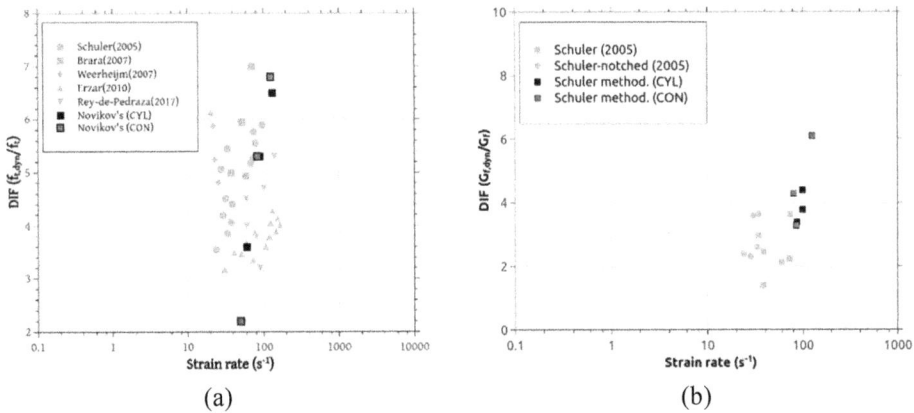

Figure 9:   Comparison between resulted DIF parameter and literature ones. (a) Tensile strength DIF; and (b) Fracture energy DIF. *(Source: Data taken from [14], [25]–[29].)*

Second, in Fig. 9(b) the analysis of different fracture energy values is shown. Due to the similarity of the experimental configuration, the only reference which is considered for the comparison is the work from Schuler et al. [25]. In this case, the comparison complement the founded data, presenting the lack of data with respect to the dynamic fracture energy analysis. As occurred with the dynamic tensile strength, the compared fracture energy shows a common trend for both studies. The critical strain rate seems to be defined at values close to 100 s^{-1}, with a sudden and exponential increase in the fracture energy reached for values of strain rate above that point.

6 CONCLUSIONS

This work presents the evaluation of two main dynamic mechanical properties of concrete: tensile strength and fracture energy. For the evaluation, two different steel projectile were used, a classical cylindrical versus a conical shaped one. The conical projectile was deisgned and validated by numerical simulations.

Dynamic tensile tests were performed using a modified SHB in its spallation configuration. A strain rate dependency was founded for both projectile types. Furhtermore a qualitative analysis based on the fracture patern reveal the convenience of using conical-projectile geometries for spalling tests. The influence and importance of the projectile's shape on the results was noted on the evolution of strains at the free end of the specimens. Therefore, the introduction of triangular pulse during the tensile loading, lead to a concentration of stresse. The numerical simulation validation allowed aswell to determine the length of the pulse, decreasing the probability of undesired pulse superpositions.

Regarding fracture energy, experiments were performed using notched specimens. As all specimens were brocken by the notch and no diffused damage or extra cracks could be found, the influence of the projectile on the results could not be determined.

ACKNOWLEDGEMENTS
The authors gratefully acknowledge the Ministerio de Ciencia, Innovación y Universidades (MCIU), Agencia Estatal de Investigación (AEI) and Fondo Europeo de Desarrollo Regional (FEDER) for providing financial support for this work under grant PGC2018-097116-A-I00.

REFERENCES
[1] Walley, S.M., The origins of the Hopkinson Bar technique. *The Kolsky-Hopkinson Bar Machine: Selected Topics*, ed. R. Othman, Springer International Publishing: Cham, pp. 1–25, 2018.
[2] Millon, O., Ruiz-Ripoll, M.L. & Hoerth, T., Analysis of the behavior of sedimentary rocks under impact loading. *Rock Mechanics and Rock Engineering*, **49**(11), pp. 4257–4272, 2016. DOI: 10.1007/s00603-016-1010-4.
[3] Birkimer, D.L. & Lindemann, R., Dynamic tensile strength of concrete materials. *Journal Proceedings*, **68**(1), pp. 47–49, 1971. DOI: 10.14359/11293.
[4] Lindholm, U.S. & Yeakley, L.M., High strain-rate testing: Tension and compression. *Experimental Mechanics*, **8**(1), pp. 1–9, 1968. DOI: 10.1007/BF02326244.
[5] Semblat, J.-F., Luong, P. & Gary, G., 3D-Hopkinson Bar: New experiments for dynamic testing on soils. *Soils and Foundations*, **39**, 2009. DOI: 10.3208/sandf.39.1.
[6] Zielinski, A.J. & Reinhardt, H.W., Stress-strain behaviour of concrete and mortar at high rates of tensile loading. *Cement and Concrete Research*, **12**(3), pp. 309–319, 1982. DOI: 10.1016/0008-8846(82)90079-5.
[7] Ožbolt, J., Weerheijm, J. & Sharma, A., Dynamic tensile resistance of concrete: Split Hopkinson bar test. *Proceedings of the 8th International Conference on Fracture Mechanics of Concrete and Concrete Structures, FraMCoS 2013*, pp. 205–216, 2013.
[8] Kolsky, H., An investigation of the mechanical properties of materials at very high rates of loading. *Proceedings of the Physical Society. Section B*, **62**(11), p. 676, 1949. DOI: 10.1088/0370-1301/62/11/302.
[9] Diamaruya, M., Kobayashi, H. & Nonaka, T., Impact tensile strength and fracture of concrete. *Le Journal de Physique IV*, **7**(C3), pp. 253–258, 1997. DOI: 10.1051/jp4:1997345.
[10] Zielinski, A.J. & Reinhardt, H.W., Fracture of concrete and mortar under uniaxial impact tensile loading. PhD thesis, Delft University Press, 1982.
[11] Ross, C.A., Thompson, P.Y. & Tedesco, J.W., Split-Hopkinson pressure-bar tests on concrete and mortar in tension and compression. *Materials Journal*, **86**(5), pp. 475–481, 1989. DOI: 10.14359/2065.
[12] Weerheijm, J., Concrete under impact tensile loading and lateral compression. Delft University, 1992.

[13] Klepaczko, J.R. & Brara, A., An experimental method for dynamic tensile testing of concrete by spalling. *International Journal of Impact Engineering*, **25**(4), pp. 387–409, 2001. DOI: 10.1016/S0734-743X(00)00050-6.

[14] Erzar, B. & Forquin, P., An experimental method to determine the tensile strength of concrete at high rates of strain. *Experimental Mechanics*, **50**(7), pp. 941–955, 2010. DOI: 10.1007/s11340-009-9284-z.

[15] Forquin, P. & Erzar, B., Dynamic fragmentation process in concrete under impact and spalling tests. *International Journal of Fracture*, **163**(1–2), pp. 193–215, 2010. DOI: 10.1007/s10704-009-9419-3.

[16] Dean, A.W., Heard, W.F., Loeffler, C.M., Martin, B.E. & Nie, X., A new Kolsky bar dynamic spall technique for brittle materials. *Journal of Dynamic Behavior of Materials*, **2**(2), pp. 246–250, 2016. DOI: 10.1007/s40870-016-0062-6.

[17] Gálvez Díaz-Rubio, F., Rodríguez Pérez, J. & Sánchez Gálvez, V., The spalling of long bars as a reliable method of measuring the dynamic tensile strength of ceramics. *International Journal of Impact Engineering*, **27**(2), pp. 161–177, 2002. DOI: 10.1016/S0734-743X(01)00039-2.

[18] Rey de Pedraza, V., Ruiz-Ripoll, M.L., Roller, C., Cendón, D.A. & Gálvez, F., Validation of two different analysis techniques to obtain dynamic mechanical properties of concrete using a modified Hopkinson Bar. *International Journal of Impact Engineering*, **161**(1), 104107, 2022. DOI: 10.1016/j.ijimpeng.2021.104107.

[19] C09 Committee, Test method for compressive strength of cylindrical concrete specimens. ASTM International.

[20] RILEM, Determination of the fracture energy of mortar and concrete by means of three-point bend tests on notched beams. 1985. DOI: 10.1007/BF02472918.

[21] Ruiz-Ripoll, M., Millon, O. & Hoerth, T., Dynamic behavior of brittle geological materials under high strain rates. *Anales de la Mecánica de la Fractura*, pp. 178–183, 2015.

[22] Mechtcherine, V., Millon, O., Butler, M. & Thoma, K., Mechanical behavior of SHCC under impact loading. *High Performance Fiber Reinforced Cement Composites 6*, eds G.J. Parra-Montesinos, H.W. Reinhardt & A.E. Naaman, Springer Netherlands: Dordrecht, pp. 297–304, 2012.

[23] Kolsky, H. & Shi, Y.Y., Fractures produced by stress pulses in glass-like solids. *Proceedings of the Physical Society*, **72**, pp. 447–453, 1958. DOI: 10.1088/0370-1328/72/3/317.

[24] Novikov, S.A., Divnov, I.I. & Ivanov, A.G., The study of fracture of steel, aluminium and copper under explosive loading. *Fizika Metallov i Metallovedeniye*, **21**, pp. 608–615, 1966.

[25] Schuler, H., Mayrhofer, C. & Thoma, K., Spall experiments for the measurement of the tensile strength and fracture energy of concrete at high strain rates. *International Journal of Impact Engineering*, **32**(10), pp. 1635–1650, 2006. DOI: 10.1016/j.ijimpeng.2005.01.010.

[26] Brara, A. & Klepaczko, J.R., Experimental characterization of concrete in dynamic tension. *Mechanics of Materials*, **38**(3), pp. 253–267, 2006. DOI: 10.1016/j.mechmat.2005.06.004.

[27] Brara, A. & Klepaczko, J.R., Fracture energy of concrete at high loading rates in tension. *International Journal of Impact Engineering*, **34**(3), pp. 424–435, 2007. DOI: 10.1016/j.ijimpeng.2005.10.004.

[28] Weerheijm, J. & Van Doormaal, J.C.A.M., Tensile failure of concrete at high loading rates: New test data on strength and fracture energy from instrumented spalling tests. *International Journal of Impact Engineering*, **34**(3), pp. 609–626, 2007. DOI: 10.1016/j.ijimpeng.2006.01.005.

[29] Rey-De-Pedraza, V., Cendón, D., Sánchez-Gálvez, V. & Gálvez, F., Measurement of fracture properties of concrete at high strain rates. *Philosophical Transactions of The Royal Society A Mathematical Physical and Engineering Sciences*, **375**, 20160174, 2017. DOI: 10.1098/rsta.2016.0174.

PRE-STRESSED REINFORCED CONCRETE ELEMENTS UNDER BLAST LOADING: NUMERICAL ANALYSIS AND SHOCK TUBE TESTING

CHRISTOPH ROLLER, MALTE VON RAMIN & ALEXANDER STOLZ
Fraunhofer Institute for High-Speed Dynamics, Ernst-Mach Institut (EMI), Germany

ABSTRACT
Advanced numerical and experimental analysis of complex structural loading conditions is presented within this paper. Major building components reinforced concrete (RC) walls are investigated with regard to their detonation resistance in various pre-stressed states. A multi-step simulation approach using successively both implicit and explicit integration schemes is followed to model the coupled static and dynamic loading. The simulation results underline the validity of the chosen modelling approach. A comparison of experimental and numerical values shows good agreement for deformation behaviour as well as for damage pattern. Beyond these predictive calculations further parameter variations indicate the dependency of highly dynamic structural response on quasi-static pre-load conditions.
Keywords: blast loading, hydrocode simulation, shock tube, pre-stressed reinforced concrete.

1 INTRODUCTION

Nowadays, an increased number of natural and manmade hazards (fire, impact, explosion [1]) needs to be considered during design phase of new buildings and requires reassessment of existing structures. Focusing on events at time range of shock phenomena, the main load bearing components have to withstand the extraordinary load conditions up to a level where fatal debris ejection is prevented and integrity as well as stability of the constructions is ensured. To cover realistic built-in stress states, analysis may not only focus on the exceptional load case itself, but include common design loads. This is the part structural engineers are quite familiar with. They deal with quasi-static loads such as live and dead loads on a day-to-day basis, even in combination with dynamic loading such as earthquake motion [2], [3]. Common approaches from structural dynamics such as response spectra [4], [5] are well established in this field and embedded in building standards [6]. In contrast, design against high-speed phenomena such as impact and explosion is rarely standardized. These phenomena rather belong to the field of natural scientists and thus analysis methods differ from common engineering practice. Besides sophisticated experimental trials, advanced numerical simulation offer the best alternative to investigate resistance of structural components under this combined static–dynamic loading in detail [7]. However, finite element calculations of structural components usually focus on single or at least similar load regimes. While common structural engineering problems can numerically be solved with implicit integration schemes, the shock problems are solved using explicit time integration. A recent study on exposed reinforced concrete (RC) walls includes a combination of both to tackle the given transient problem. The numerical analysis and related trials are described briefly in the following.

2 SIMULATION APPROACH AND VALIDATION

The investigations conducted within the ANSYS Workbench environment comprise four different initial load conditions with three dynamic load regimes each. Besides purely vertical, purely horizontal and combined vertical plus horizontal pre-load, configurations without any pre-load are investigated. The latter serve as reference to classify the influence

WIT Transactions on The Built Environment, Vol 209, © 2022 WIT Press
www.witpress.com, ISSN 1743-3509 (on-line)
doi:10.2495/HPSU220151

of the single pre-load configurations on the structural response of the exposed RC-walls. Table 1 gives an overview about the scope of the investigations.

Table 1: Scope of investigations.

Configuration	Pre-load		Dynamic load
	Vertical – n0	Horizontal – p0	DIN 13123-1 [8]
n0	√	x	EPR1
	√	x	EPR2
	√	x	EPR3
n0p0	√	√	EPR1
	√	√	EPR2
	√	√	EPR3
p0	x	√	EPR1
	x	√	EPR2
	x	√	EPR3
0	x	x	EPR1
	x	x	EPR2
	x	x	EPR3

The modelled RC-walls measure 186 cm in height, 60 cm in width and 12.5 cm over thickness. Concrete body as well as reinforcement, consisting of steel bars and shear links, are individually discretized according to their dimensions in full-scale (see Fig. 1). Supports at top and bottom of the bodies are modelled using translational and rotational boundary conditions. Pre-load conditions are applied to the corresponding surfaces via appropriate stress boundary conditions, too. Gauge points are defined at significant locations at mid-span and on top of the concrete body for analysis of displacement and stress histories.

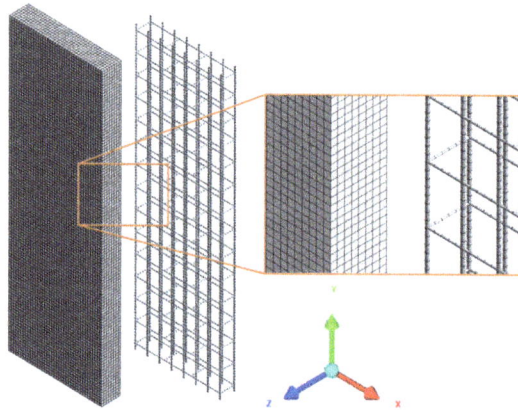

Figure 1: Numerical model of RC-wall with discrete reinforcement.

2.1 Combined loading–combined solvers

Analysis of such walls under combined quasi-static and highly dynamic loading requires more than one time integration scheme to produce efficient simulations. Thus, a coupled multi-step simulation approach using both implicit and explicit solvers is followed for the present transient problem.

Based on the developed model, first the quasi-static load case is computed as a pre-stressed analysis using the ANSYS Mechanical APDL solver. Simulation results such as vertical and horizontal deformation plus vertical stress are identified for matching of data with next steps. Second, the data including geometry model, mesh size and analysis settings is transferred to the explicit dynamics module. The information from pre-stress analysis is applied via displacement-orientated initial condition to position the explicit nodes in a pre-defined time. During the initialization strains and stresses are computed by the ANSYS Explicit STR solver as usual.

To change the calculation from a dynamic solution to a relaxation iteration which converges to a state of stress equilibrium a static damping constant R is specified. Depending on the average time step ts and the longest period of the system's motion T constant R can be chosen from eqn (1) for ratios of $ts << T$:

$$R = 2 \cdot \left(\frac{ts}{T} \right) \qquad (1)$$

Thus, an intermediate step for determination of T is integrated into the process. The Modal module is linked to the static structural system transferring all set-up and solution data from pre-stress analysis. Running the modal analysis gives the required first natural period. Taking into account an average value for ts from a preliminary explicit calculation without static damping R amounts to a value of 2.6710^{-4} ms in this case which is within the recommended range.

The third step includes the transfer of the computed data to the hydrocode ANSYS AUTODYN [9]. Here, a change of material models is required. Although the explicit dynamics solver from the previous step uses the same explicit solver as hydrocode, the material data is linked to the static structural module and thus the solver refers to "implicit models". These models are not designed for transient problems in the range of shock loading. Thus, well established material models that cover phenomena as compaction and strain rate effects are applied for concrete and steel, namely RHT [10] and Johnson–Cook [11].

To discuss the intended equilibrium state after pre-load initiation, Fig. 2 exemplarily shows the displacement-time history for the top face centre of the RC-wall under combined pre-load determined by the single simulation steps. Comparing results from implicit and explicit pre-simulation, it is apparent that the explicit curve is neither balanced nor does it match with the static solution. Taking into account the hydrocode calculation with implicit material models, it is clear that this mismatch cannot be related to the different solvers, since this curve does agree well with the implicit solution. It is assumed that this difference results from an insufficient bonding between steel beam elements and concrete volume body in the explicit dynamics module. Merging the nodes is realized during implicit simulation set-up and via the corresponding option for all unstructured nodes within the hydrocode environment, but both is not an option for the explicit dynamics module. Since the focus is on a coupled simulation using the hydrocode, this short coming is neglected.

However, comparing solutions from the hydrocode the expected influence resulting from different material models is observed. To achieve the intended initial displacement the applied pre-load is linearly reduced in the hydrocode. The resulting curve approves this approach in matching the targeted implicit curve.

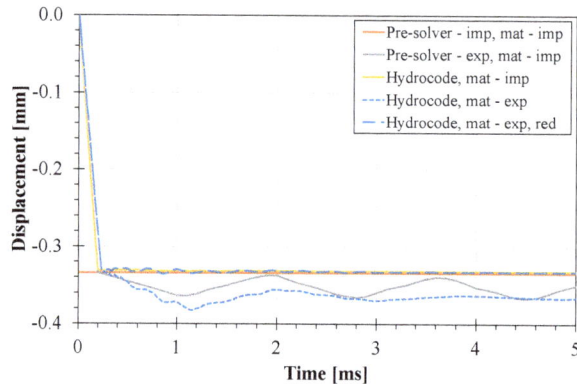

Figure 2:   Evolution of vertical displacement at top of RC-wall in order to achieve equilibrium state.

2.2 Validation of approach

Results of shock tube tests are used to validate the simulation approach with regard to chosen discretization, boundary conditions and material models. The trials are conducted using the shock tube BlastStar [12] at Fraunhofer EMI facilities. Before being loaded with air pressure shock wave the RC-specimen have been pre-stressed perpendicular to as well as along the plate axis. To apply quasi-static bending load perpendicular to the plate axis, the shock tube has been enhanced by installation of a ventilation adapter enabling over-pressure respectively low-pressure regimes. Comprehensive computation of structural and gas dynamics for ventilator and its mounting have been conducted with regard to the dynamic shock wave loading. To apply quasi-static compression load along the plate axis as well as to realize sufficient support conditions further design work has been necessary. Fig. 3 illustrates the developed supporting frame including dowel pins and pre-stressed rods in ways of technical drawing and final built-in state. Using these extra features, it is possible to apply pre-loads up to $p = 30$ kN/m^2 in horizontal and $N = 2,400$ kN in vertical direction.

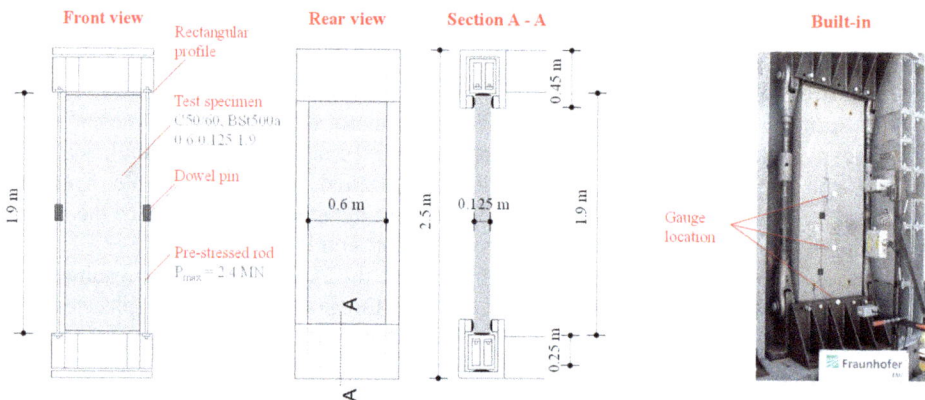

Figure 3:   Test set-up for RC-walls under combined loading as construction detail (left) and in built-in state (right).

Main focus of the investigations is on deformation behaviour of the RC-wall, since the flexural strength is the decisive parameter for walls loaded by air shock pressure related to a far range detonation. Accordingly, an extensive instrumentation has been provided. Apart from pressure sensors for measurement of applied dynamic loading on the exposed face, laser and high-speed video equipment plus strain gauges have been installed on the opposite side for specification of the structural response.

First, maximum deformation of the single test series is observed. Fig. 4 contains the pair of values from numerical and experimental data source. The red diagonal represents the line where results of simulation and experiment are in conformity. The blue marks displayed in the diagram are located very close to that line for all types of pre-loading. This indicates that the particular results are in good correlation. Hence, the simulations are suitable for further analysis of the deformation behaviour.

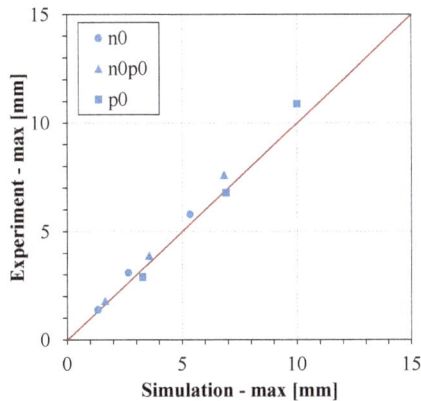

Figure 4: Comparison of numerically and experimentally determined maximum deflections.

A second significant parameter being used for validation is the emerging degree of damage. Most analysed configurations stayed in the damage-free elastic range. Only the configuration without any pre-load showed some crack initiation on the non-loaded side at increased dynamic loading due to tensile loading. Fig. 5 exemplarily compares damage pattern of the RC-elements after EPR3 loading [8]. The colour coding of the simulation results represents a strength related damage. Blue equals to intact concrete regions, whereas red represents regions with no residual strength left, meaning completely damaged regions. During previous studies it has been seen that green–yellow colour-coding is related with regions of crack formation. In the light of the above a good correlation between simulation and experiment is given here, too: Various horizontal cracks can be detected in mid-span. Based on these and the above mentioned conformity in results, it follows that the simulations provide sufficient predictive quality posse the ability of prognosis and can be used for further analysis.

3 INFLUENCE OF SINGLE LOAD LEVELS

To begin with, the computed first deflection for each static-dynamic loading is comparatively displayed in Fig. 6 for blast load EPR2 [8]. The reducing effect for the single pre-load conditions is clearly given. It is shown in particular that the deflection curve resulting from

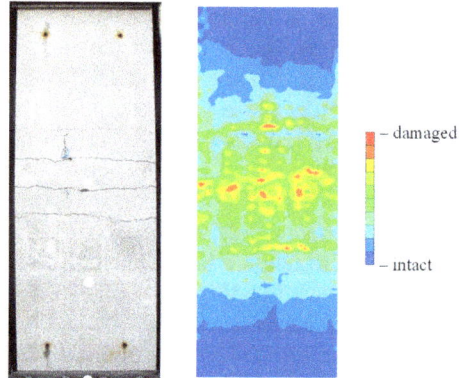

Figure 5: Degree of damage of the RC-plates after EPR3 [8].

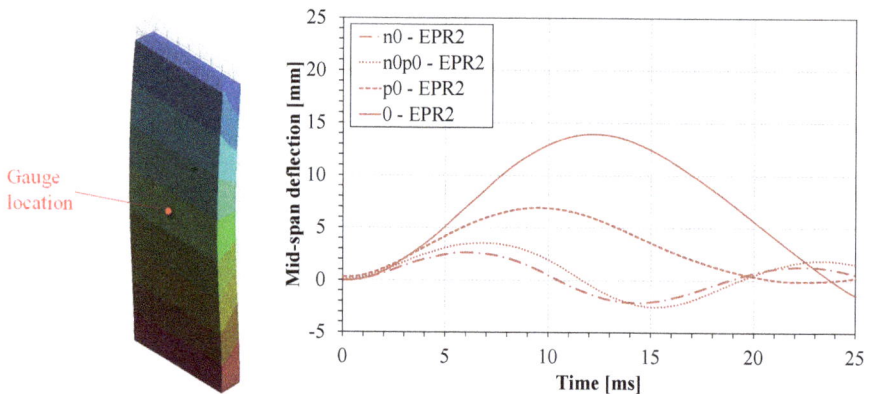

Figure 6:   Comparison of numerically determined displacement-time histories for EPR2 [8] and gauge location.

combined pre-load cannot be extrapolated by superposition of the curves resulting from single pre-loads but that it lies in between them. More precisely, the behaviour due to combined pre-load is closer to the one from purely vertical pre-load. This phenomenon similarly appears for the configurations with decreased and increased blast loads. However, it has not been theoretically fully understood and needs to be studied further.

Focusing on the effect of the single pre-loads individually a number of parameter variations have been computed with different levels of pre-load and blast load. This also demonstrates the added value by using simulations, since any experimental parameter studies would consume a multiple amount of time and money.

3.1 Level of vertical pre-load

Starting from the vertical static compressive stress referred to above, the stress has been divided by two in a first variation and doubled in a second variation for each of the three dynamic explosion pressure classes. The result of the numerical investigation is given in terms of displacement-time histories (see Fig. 7 (left)). The previously noted effect can be confirmed by the parameter study: Increased vertical pre-load leads to decreased maximum deflection and vice versa. Additionally it is noted that the varying boundary conditions influence the natural oscillation behaviour: Increased vertical pre-load leads to decreased natural period (increased natural frequency) and vice versa.

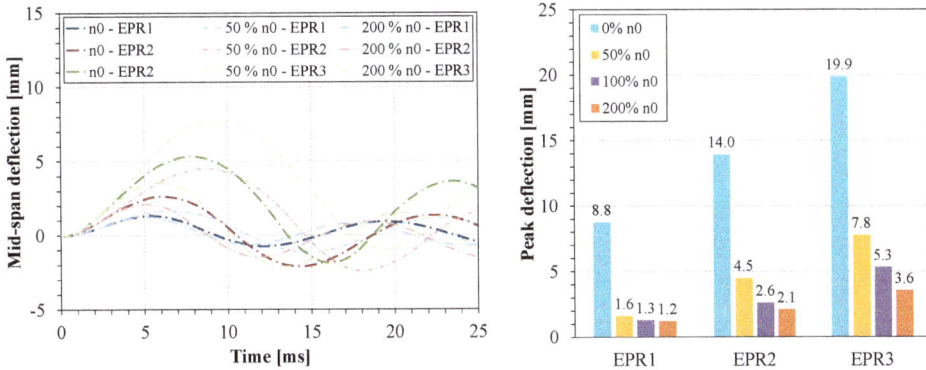

Figure 7:   Comparison of numerically determined displacement-time histories for varying vertical pre-loads (left) and blast levels and of absolute maximum deflections with and without vertical pre-load variations for single dynamic blast levels (right).

To specify the effect further, the determined maximum deflections are compared with the ones from reference simulations without any pre-load. The comparison of absolute values in Fig. 7 (right) illustrates the significant influence of the vertical pre-load. At a dynamic load of 100 kPa (EPR2) for example the amplitude can be reduced from 14.0 mm down to 2.1 mm at highest vertical pre-load. This equals to a reduction down to 15% related to the reference configuration. In addition it is found that the degree of reduction also depends on the level of blast load: The mitigating effect decreases at increasing blast level. Generally a reduction between 13% and 39% is numerically detected.

It should be noted that the promising simulation results of decreased deformation have to be evaluated with care: Due to the dynamically induced bending deformation the initially axial pre-load becomes an eccentrically vertical load at increased bending stress. Thus, the risk of flexural buckling needs to be considered in building practice relevant design for combined static vertical and dynamic horizontal load. There is still research needs in this field.

3.2 Level of horizontal pre-load

In analogy to the previous variant analysis, for the horizontal pre-load the planar static compressive stress is divided by two on the one hand and doubled on the other hand, too. The result of the simulation runs is illustrated in way of displacement-time histories again, see Fig. 8 (left). In contrast to the situation of vertical pre-load it is not possible to confirm a steady effect. A decreased horizontal pre-load does lead to an increased maximum deflection, but so does an increased horizontal pre-load. It seems that a limited effect is acting here which is valid for a certain range of parameters only.

A significant influence on the natural period cannot be detected here and thus is assumed to be negligible for this parameter variation.

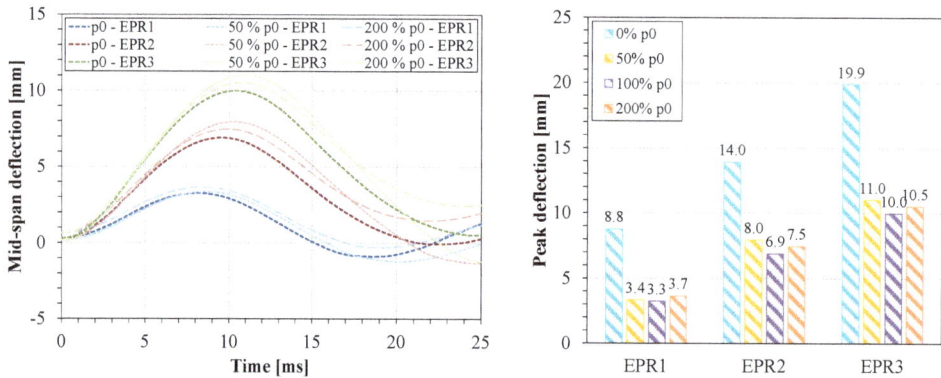

Figure 8:   Comparison of numerically determined displacement-time histories for varying horizontal pre-loads and blast levels and of absolute maximum deflections with and without horizontal pre-load variations for single dynamic blast levels (right).

The absolute values of the computed maximum deflections are compared to the values from the reference simulation without any pre-load in Fig. 8 (right). On the one hand the diagram clearly illustrates the existing influence of the horizontal pre-load. On the other hand, the increased deflection is evidenced for both half and double horizontal pre-load. This is an effect that needs to be investigated further. A comparison with Fig. 8 indicates that the applied vertical pre-load affects the maximum deflection to a greater extend. Whereas a reduction between 13% and 39% is detected for the previous pre-load study, here values between 37% and 57% are noted. Additionally, it is found that the degree of reduction does not depend on the level of blast load for the horizontal pre-load: The mitigating effect becomes rather constant at increasing blast level.

4 CONCLUSIONS

Based on the parametric studies new findings are introduced in the field of pre-loaded protective structures. The combined experimental and numerical investigations outline the importance of considering initial stress states when dealing with blast-resistant design: Both vertical and horizontal pre-load significantly influence the structural response of exposed RC-walls. This includes but goes beyond characteristics such as peak deformation, oscillation and residual load capacity.

The coupled simulation approach provides realistic results for the investigated reinforced concrete members under complex loading conditions for both, deformation behaviour and damage pattern. Thus, the approach can be used for further investigations on this topic.

REFERENCES

[1]   Ibrahimbegovic, A., *Extreme Man-Made and Natural Hazards in Dynamics of Structures*, Springer: Dordrecht, Netherlands, 2007.
[2]   Bachmann, H., *Erdbebensicherung von Bauwerken*, 2, überarbeitete Auflage. Birkhäuser Verlag: Basel, Switzerland, 2002.

[3]   Butenweg, C. & Roeser, W., *Erdbebenbemessung von Stahlbetontragwerken nach DIN EN 1998-1*, Goris & Hegger, eds, Stahlbetonbau aktuell 2012 - Praxishandbuch, Bauwerk Verlag: Beuth, Berlin, Germany, 2012.

[4]   Newmark, N.M. & Hall, W.J., *Earthquake Spectra and Design*, Berkeley, CA, USA, 1982.

[5]   Chopra, A.K., *Dynamics of Structures*, 2nd ed., Prentice Hall: Upper Saddle River, NJ, USA, 2001.

[6]   Deutsches Institut für Normung, DIN 4149:2005-04: Buildings in German earthquake areas – Design loads, analysis and structural design of buildings, Beuth, Berlin, 2005.

[7]   Riedel, W. & Mayrhofer, C., Customized calculation methods for explosion effects on structural building components. *Int. Symp. on Structures under Earthquake, Impact and Blast Loading – IB'08*, Osaka University, Arata Hall, Osaka, Japan, 2008.

[8]   Deutsches Institut für Normung, DIN EN 13123-1: Windows, doors and shutters – Explosion resistance – Requirements and classification – Part 1: Shock tube, Beuth, Berlin, 2001.

[9]   Century Dynamics Inc., Autodyn: Interactive non-linear dynamic analysis software – Theory Manual 4.0, Horsham, UK, 1998.

[10]   Riedel, W., Beton unter dynamischen Lasten, Meso- und makromechanische Modelle und ihre Parameter, Schriftenreihe ε – Forschungsergebnisse aus der Kurzzeitdynamik, Fraunhofer IRB Verlag: Heft 5, 2004.

[11]   Johnson, G.R. & Cook, W.H., Fracture characteristics of three metals subjected to various strains, strain rates, temperatures and pressures. *Engineering Fracture Mechanics*, **21**(1), pp. 31–48, 1985.

[12]   Klomfass, A., Kranzer, C., Mayrhofer, C. & Stolz, A., A large new shock tube with square test section for the simulation of blast events. *Proceedings of the 22nd MABS – Military Aspects of Blast and Shock*, Bourges, France, 4–9 Nov., 2012.

COLLAPSE FRAGILITY CURVES FOR SEISMIC ASSESSMENT OF SUPERPLASTIC SHAPE MEMORY ALLOY IN REINFORCED CONCRETE STRUCTURES

FARAH JAAFAR & GEORGE SAAD
Department of Civil and Environmental Engineering, American University of Beirut, Lebanon

ABSTRACT

Contemporary building regulations intend to define the standards for design and construction while contemplating safety and serviceability for the occupants. Even though these codes safeguard occupants' lives under severe earthquakes, damage will occur, inducing stiff repairs and in certain cases building demolition. To address this issue, the design of buildings in seismic regions should aim to be more resilient structures that sustain little or no damage when subjected to extreme loading conditions. This study investigates the use of super-elastic shape memory alloys (SSMA) as partial replacement of steel reinforcement in reinforced concrete (RC) structures to enhance their seismic performance. SSMA is considered a particular type of smart alloys that has the ability to undergo large deformations and return to its original shape after the application of a reverse load, and hence can enhance the performance envelope of the structure. A sensitivity analysis is conducted to assess the efficiency of using SSMA at different locations in reinforced concrete frames. Fragility curves evaluating the seismic performance of an eight-story RC frame reinforced with steel and SSMA at different locations are developed. The results reveal the efficient competence of SSMA reinforced structures at different performance levels as they need greater forces to reach their plastic limit, hence increasing the overall performance of the structure.

Keywords: SSMA, performance-based design, earthquake, fragility curves.

1 INTRODUCTION

Smart materials (SM) have become topics of interest in recent research studies due to their ability to adaptively respond to external changes [1]. Depending on their type, these materials can be triggered by a change in temperature, stress or magnetic/electric field and respond by a change in their composition or properties [2]. The main advantages of SM consist of their high mechanical performance, high damping capacity, large actuation force, compactness, and lightness. Whereas SM most challenging and critical disadvantages are the high cost and environmental dependencies of these materials [3].

Shape memory alloy (SMA) is a prominent type of smart materials that have found extensive use in the engineering sector due to its two unique nonlinear phenomena, the shape memory effect (SME) and super-elasticity. The first class of SMA (SME) guarantees the recovery of large mechanical strains by heating the material above a critical temperature. On the other hand, super-elastic SMA (SSMA) has the capability to undergo large deformations and return to its original configuration upon removal of externally applied loads, without changing the ambient temperature of the system. These alloys find their applications in civil engineering due to their capability to absorb strain energy without durable damage and to withstand fatigue resistance under wide strain cycles [4]. These characteristics of SMA make it a tempting material to be used in concrete structures, especially in seismic regions. Earthquakes results in arbitrary motions, produce reversed cyclic loading on the structure, and hence can cause a compressive failure and tensile yielding of the concrete and reinforcing steel respectively [5]. Besides, large drifts may occur at the story level which leads to an increase in the structure's stiffness and a decrease in the serviceability level. Therefore, it is

WIT Transactions on The Built Environment, Vol 209, © 2022 WIT Press
www.witpress.com, ISSN 1743-3509 (on-line)
doi:10.2495/HPSU220161

a must to find an optimum design for structures to resist seismic lateral loads with minimal additional cost.

Different types of SMA have been used in concrete structures as external reinforcement (bars and rods), partial replacement of steel reinforcement (bars), for strengthening and retrofitting (wires and plates) or as fibers embedded in the cementitious composite [6]. Previous studies have experimentally outlined the relevance of SME in confining concrete elements, where an increase in the loading capacity and a decrease in the permanent deformation was noted due to the recovery stresses imposed when the SMA was heated [7]. Despite the advantages of many types of SMA, super-elastic nickel-titanium based SMA was found to be the most suitable and durable for practical use in active systems, where no external factors are needed to activate the super-elasticity effect [8]. For instance, Khaloo et al. investigated the effect of the ratio of partial reinforcement replacement by SSMA rebars on the behavior of a cantilevered reinforced concrete beam under lateral loading [9]. The study showed that SSMA material is capable of recovering its initial state, in addition to producing tensile forces which are responsible for cracks closure. Abdulridha et al. have conducted an experimental application to study the performance of a simply supported beam with SSMA bars under different loading conditions [10]. The results revealed the advantages of SSMA over regular steel in recovering large plastic deformations upon removing of loading, and hence the closure of cracks.

Many researchers have used SSMA at the plastic hinge location in different concrete elements subjected to cyclic loading to demonstrate the efficiency of this material in recovering post-yield deformations [11]. For example, Alam et al. investigated the application of SSMA in the plastic hinge area of a beam–column connection throughout a numerical analysis [12]. Their results demonstrated the superiority of SSMA–RC connection over regular steel–RC connection due the recentering capability of such material even for large deformations. Furthermore, Nahar et al. examined numerically the dynamic performance of concrete beam–columns joints reinforced with different SSMA types, at the plastic hinge location, under non-linear static pushover and reversed cycling loading. A satisfactory energy dissipation capacity and minimal residual strains were reported, which induce the least maintenance and rehabilitation cost after the post-earthquake deformation [13].

In another study, Alam et al. assessed the seismic behavior of an eight-story RC frame reinforced with SSMA along the plastic hinge length of the beams, under ten ground motion excitations, as compared to regular steel–RC frame [14]. The results showed the superiority of SSMA in reducing both inter-story and top story residual drift.

In the previously mentioned studies, the SSMA usage was confined to specific elements under specific loading conditions. Although the results demonstrated the importance of SSMA, they did not consider the probability of building damage when the location of SSMA or the intensity of loading change. Under seismic excitations, the design of buildings is probabilistic rather than deterministic, hence the importance of determining the probability of structural failure when any of the design parameters (location and length of SSMA and ground motion intensities) change. This paper addresses the missing segments in the literature by developing collapse fragility curves for RC buildings with different design parameters. This study starts by designing an eight-story steel–reinforced concrete frame (Frame 1) in a specific seismic zone. This design is replicated for two additional RC frames, where one (Frame 2) has SSMA rebars along the columns of the first floor and the second (Frame 3) is fully reinforced with SSMA along all its columns. The three frames are subjected to 21 different ground motion records and are modeled using the Open System for Earthquake Engineering Simulation (OpenSees) [15]. The collapse fragility curves are developed for the three RC frames under increasing earthquake intensities and for different damage states. The purpose of this paper is to find the earthquake intensity margin where the performance of SSMA reinforced frames outstand that of regular steel–reinforced frame.

2 REINFORCED CONCRETE FRAME CHARACTERISTICS

The frame to be considered for analysis is designed in accordance with ACI 318-19 [16] provisions for element design, and ASCE 7-16 [17] regulations for load combination and seismic design; and it is assumed to be located in a moderate seismic zone. In compliance with the International Building Code (IBC) [18], the considered ground parameters are associated with a 475-year return period, which equates to Zone 2B [19].

2.1 Frame model

A medium rise eight-story, four bay RC frame is considered in this study, as shown in Fig. 1. The frame has a typical floor height of 4 m and span length of 5 m. The preliminary design of the frame is performed using Structural Analysis Program 2000 [20] presuming a 2D planar model and base fixation. Assuming that frame is only subjected to its own weight (D) and seismic excitation (E), the beams and columns are designed for the following load combination:

$$0.9D \pm 1E. \tag{1}$$

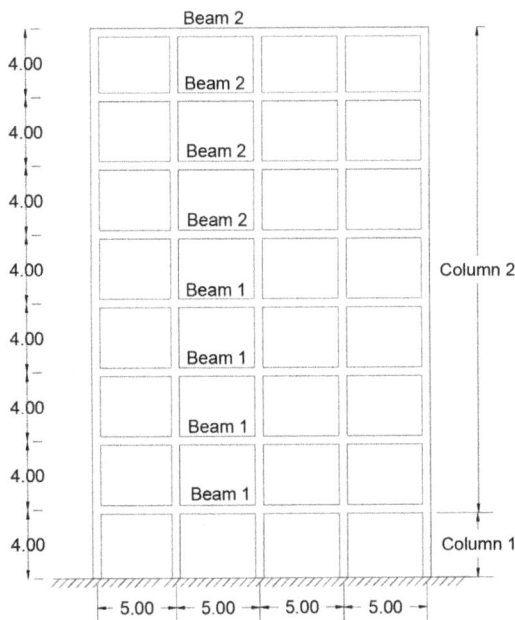

Figure 1: Eight story RC frame elevation.

The frame is designed such that it is fully reinforced with steel rebars, and its peak response is obtained from a response spectrum analysis. The response spectrum function is defined by IBC, having the following parameters:

S_s: Spectral acceleration at 0.2 s = 1.2 g

S_1: Spectral acceleration at 1 s = 0.4 g

The design response spectrum adopted is presented in Fig. 2, where T* is the first natural period of the steel reinforced concrete frame. It is evident to claim that in pursuance of a seismic analysis, the strong column–weak beam design is adopted; hence the columns are the significant elements. The beams and columns sections, in addition to the reinforcement ratio obtained from the spectral analysis are used to model the frame in Opensees. It is to be noted that only the steel reinforced frame is thoroughly designed in this study, while for the other SSMA reinforced frames, the steel rebars are simply replaced by SSMA material at certain locations.

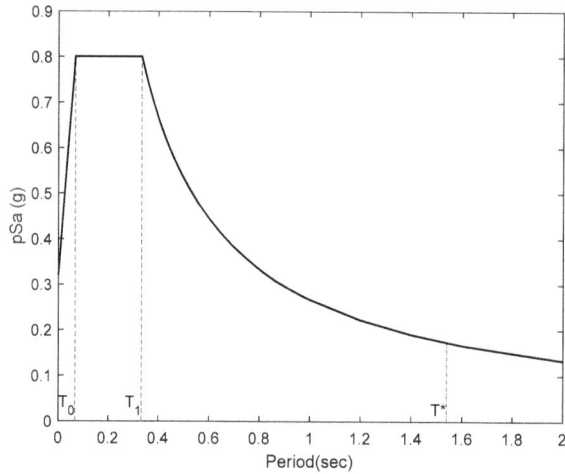

Figure 2: Design response spectrum.

2.2 Material properties

The detailed section design of Frame 1 and the material properties used in Opensees are shown in Fig. 3 and Table 1, respectively. The steel rebars are defined by Menegotto and Pinto [21] isotropic hardening material, and are assumed to be of a Grade 60, having a yielding strength of 420 MPa [22]. The SSMA bars are modeled as uniaxial self-centering material, having a flag shaped hysteretic response as shown in Fig. 4 [15]. As illustrated in Fig. 4, SSMA remains linear-elastic until the activation stress is reached, which is considered as the yielding stress in regular steel bars. Increasing the load above this limit, will cause the material to deform and plastic deformations to develop. As long as the plastic strain is less than or equal to the ultimate strain of SSMA, the material will return to its original configuration with zero residual strain upon unloading. The confined and unconfined concrete are modeled using the constitutive relationship proposed by Popovics [23] and Karsan and Jirsa [24]. The beams and columns elements are modeled using the force-beam elements (FBE), that grants the spread of plasticity along the length of the element. The convergence of FBE depends on the number of integration points along the element length, which an average of five integration points was found to be sufficient to approach the exact solution [25]. The beam and column sections are defined as fiber sections.

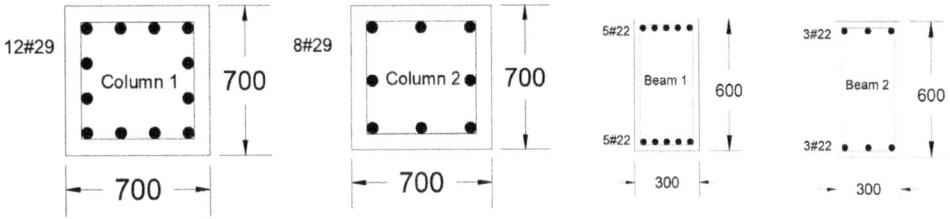

Figure 3: Columns and beams reinforcement details.

Table 1: Material properties.

Unconfined concrete	f_c (Mpa)	30
	ε_y	0.002
	ε_u	0.004
Confined concrete	f_{cc} (Mpa)	35.8
	ε_y	0.004
	ε_u	0.026
Steel rebars	E (Mpa)	200,000
	f_y (Mpa)	480
	εu	0.0051
SSMA rebars	K_1 (Mpa)	68,200
	σ_{act} (Mpa)	480
	K_2 (Mpa)	954.2
	ε_u	0.62

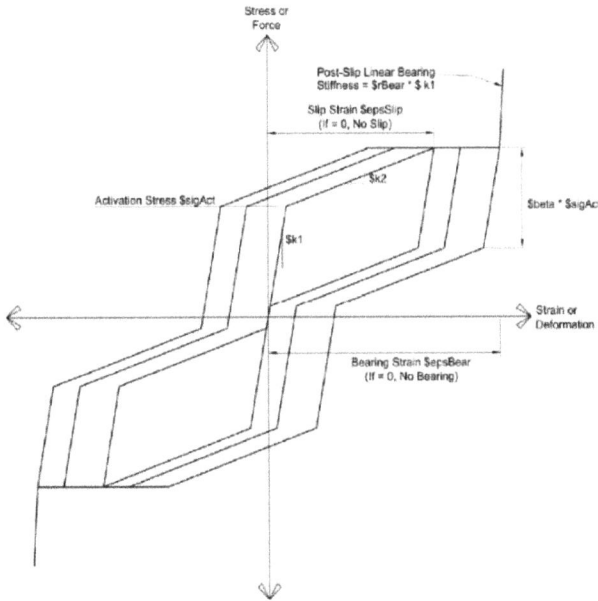

Figure 4: SSMA flag-shaped model [15].

3 NUMERICAL ANALYSES

The custom placement of SSMA is due to its high initial cost and the large deformations that can be displayed due to its low elastic stiffness. Accordingly, a dynamic uniform sine wave ground motion, with an amplitude of 4 g, is applied to the steel RC frame (Frame 1) for the sake of determining the preliminary location of SSMA. The amplitude of the ground excitation is determined such that plastic hinges start to form in the frames' vertical elements. Fig. 5 delineates the sequence of plastic hinge formation in the steel RC-frame, due to the applied ground motion. As displayed, the columns at the base were the first to deform, which is in agreement with the definition of ground shaking mechanism. Before proceeding forward, an RC frame having SSMA reinforcing bars in the columns of the first floor is subjected to the same sinusoidal excitation, to monitor the change in the plastic hinge pattern. Although the base columns are the ones to attract large forces, the plastic hinges shifted one story upward, as shown in Fig. 6. This variation of the deformation sequence is related to the different definitions of plastic hinge for both steel and SSMA reinforcement material. For instance, a plastic hinge is defined at the maximum elastic recoverable strain. In case of steel reinforcement, the latter value is nothing but the yielding strain, which is 0.0024. However, SSMA can undergo larger deformation, up to 6% strain before starting to accumulate plastic strains. Even though SSMA has a small initial stiffness as compared to steel, the large recoverable strain difference between both materials is the cause of the plastic hinge shifting.

Considering the shift of plasticity in the frame after including SSMA reinforcing bars in the first level, it may be of interest to inspect the performance of different reinforcing configurations and comparing their behaviour under seismic excitations.

Frame 1 is the originally designed steel RC frame, considered as a benchmark for comparison with SSMA–RC frames. For Frame 2, the columns reinforcing bars at the first level are replaced with SSMA bars, while all the remaining elements are reinforced with regular steel. It is veracious to say that at high earthquake intensities plastic hinges may start to form at different locations along the same member, perhaps at both ends of the base columns. To this end and to account for the all the possibilities, the assumption of having SSMA along the entire length of the column was adopted in Frame 3. It is noteworthy that these different configurations do not assume the best location of SSMA but consider the most efficient performance of the overall RC frame.

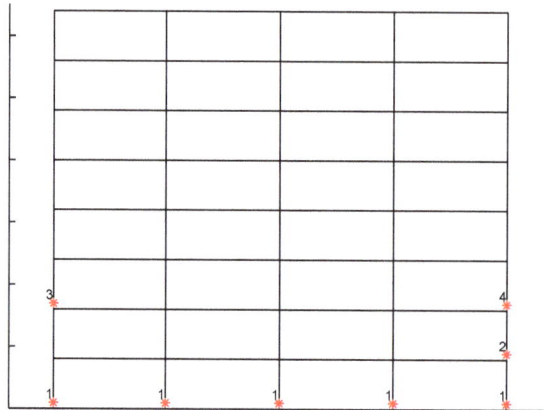

Figure 5: Plastic hinge formation of Frame 1.

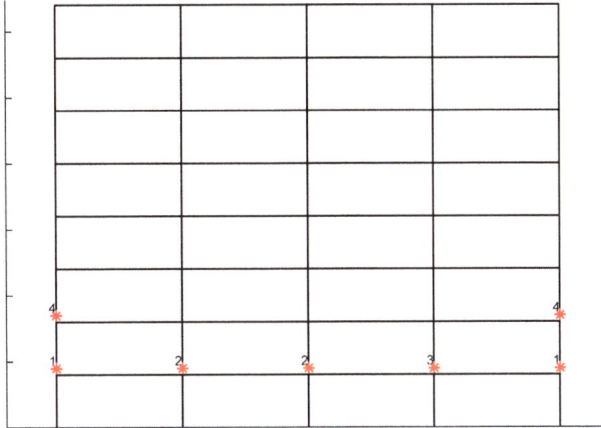

Figure 6: Plastic hinge formation of Frame 2.

It is to be noted that an eigen analysis was employed to determine the structural period for the steel RC frame (Frame 1), resulting in an average first mode period of $T_1 = 1.54$ s. The fundamental periods of Frames 2 and 3 are 1.56 and 1.57 respectively, which are slightly larger than that of Frame 1, as SSMA has a low stiffness compared to steel. However, for the ease of computation, a constant period is considered in this analysis. This assumption has no significant impact on the analysis since the spectral accelerations corresponding to three frames are almost identical, for a value of 0.173 g, 0.171 g and 0.1699 g respectively.

4 FRAGILITY ANALYSIS

The limited use of SSMA in the construction industry, despite its proven efficiency through numerical analysis, is associated with the high uncertainties of the global performance of the buildings reinforced with such material. According to the projects and research presented by the Pacific Earthquake Engineering Research (PEER), the seismic design of structures is probabilistic rather than deterministic. This is due to the seismic vulnerability induced by the characteristics of the structure, local site-effects, and earthquake intensity/frequency. Collapse fragility curves are widely known as the primary aspects in evaluating buildings performance. These curves aim to define a cumulative distribution function (CDF), which relates the ground motion intensity to the probability of structural failure to meet a certain response level.

In this section, the fragility functions will be developed from an incremental dynamic analysis. All three frames will be subjected to a set of 21 different ground motions scaled incrementally from a spectral acceleration (Sa(T$_1$)) 0.1g to 2g. The scale range was selected such that the frames subjected to each ground excitations fail all the performance levels stated in Table 2. The fragility function is defined by a lognormal cumulative distribution function:

$$P(C|\ IM{=}x) = \phi\left(\frac{\ln\left(\frac{x}{\theta}\right)}{\beta}\right), \tag{2}$$

where IM is the intensity measure adopted in the analysis, ϕ () is the standard normal cumulative function, θ is the median of the fragility function and β is the dispersion of IM. θ and β are the main parameters that define the fragility function. The procedures considered in developing the fragility curves are stated below.

Table 2: Structural performance levels.

		Structural performance levels		
		Immediate occupancy	Life safety	Collapse prevention
Steel reinforced frame	Damage	Light	Moderate	Severe
	Structure	Concrete cover is not allowed to crush in any member	Confined concrete stress $<f_{cc}$ for all elements	Confined concrete strain in columns < cracking strain
		No yielding of steel reinforcement	Maximum steel reinforcement strain $<\varepsilon y$	Maximum steel strain< 0.0045
	Drift	Maximum drift 1%	Maximum drift 2%	Maximum drift 4%
		No significant residual drift	Maximum residual drift 1%	Maximum residual drift 4%

4.1 Define limit states

In this analysis, the probability of failure is taken for three damage states: immediate occupancy (IO), life safety (LS) and collapse prevention (CP); as per FEMA-356 [26]. All three limits account for maximum deformation, residual drift, reinforcement yielding and concrete crushing, as shown in Table 2.

It is worth mentioning that the reinforcing bars are evaluated at the stress/strain yielding limit in case of steel material and at the unrecoverable strain limit in case of SSMA.

This approach of defining the limit state is likely conservative because it assumes that when the failure limit is exceeded in one element, it triggers failure of the entire structure. In many cases, gravity loads can be redistributed to nearby elements, and the axial failure of a single column will not cause complete collapse of the frame. However, we are adopting this strategy because of physiological and psychological purposes. For example, if the maximum displacement in a story exceeds three times the allowable limit, the structure may not fail but the occupants may feel dizziness because of the large sway.

4.2 Select record set

Padgett and Desroches [27] and Asgarian et al. [28] have highlighted the effect of earthquake characteristics on the overall nonlinear performance of structural systems due to the high uncertainties provided by the random nature of these ground motions. On that account, 21 ground motions were selected with a variability in terms of magnitude, rupture fault distance, D5-95 and Arias intensity. A summary of the characteristics of the ground motions is presented in Table 3.

The considered earthquake records belong to the far field set, from Pacific Earthquake Engineering Research Center [29] strong motion database, and they are chosen such that their mean coincide with the target spectrum, as shown in Fig. 7. This set covers a wide range of earthquake properties such as frequency, ground motion intensities and duration.

4.3 Normalizing record set

The ground motions are first normalized in order to remove excessive variability between records due to the dissimilarities of the properties, in terms of magnitude, distance to source,

Table 3: Ground motions properties.

Record sequence number	Earthquake name	Magnitude	PGA (m/s²)	Arias intensity (m/s)	5%–95% duration (s)	Normalization factor
31	Parkfield	6.19	0.272	0.4	13.1	1.2
132	Friuli Italy-02	5.91	0.212	0.4	4.6	1.28
136	Santa Barbara	5.92	0.202	0.2	7.5	1.00
162	Imperial Valley-06	6.53	0.204	0.9	14.8	0.935
204	Imperial Valley-07	5.01	0.274	0.3	6.5	0.795
208	Imperial Valley-07	5.01	0.255	0.1	7	1.002
233	Mammoth Lakes-02	5.69	0.183	0.2	7.7	1.11
236	Mammoth Lakes-03	5.91	0.233	0.4	6.3	0.907
248	Mammoth Lakes-06	5.94	0.314	0.5	6.8	1.0
249	Mammoth Lakes-06	5.94	0.377	1	5.1	0.959
391	Coalinga-03	5.38	0.199	0.2	14.8	1.288
406	Coalinga-05	5.77	0.519	0.8	8.5	0.742
408	Coalinga-05	5.77	0.193	0.3	8.5	0.997
409	Coalinga-05	5.77	0.216	0.3	8.5	0.979
410	Coalinga-05	5.77	0.309	0.6	6.9	0.852
413	Coalinga-05	5.77	0.228	0.5	6.5	1.06
456	Morgan Hill	6.19	0.213	0.2	16.6	1.131
457	Morgan Hill	6.19	0.201	0.3	20.4	1.107
502	Mt. Lewis	5.6	0.149	0.2	9.5	0.926
706	Whittier Narrows-01	5.99	0.235	0.7	9.1	1.078
714	Whittier Narrows-02	5.27	0.319	0.4	6.2	0.951

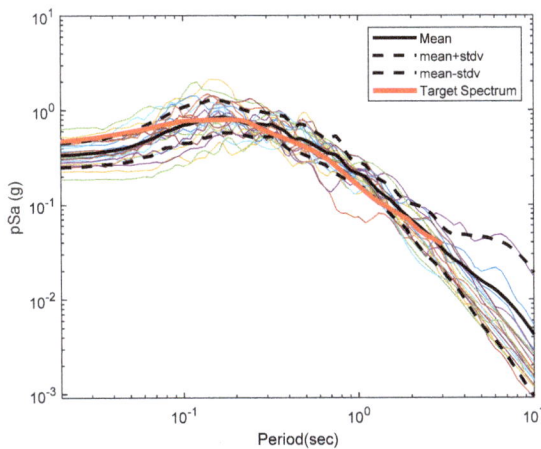

Figure 7: PEER ground motions-response spectra.

source type and site conditions, while still maintaining the record-to-record variability necessary for accurately predicting collapse fragility. The normalization is done with respect to the peak ground velocity (PGV) values. For any ground motion in the set, the normalization factor of both horizontal components is given by

$$CF_i = \frac{median(PGV_{PEER,i})}{PGV_{PEER,i}},$$ (3)

where $PGV_{PEER,i}$ is the geometric mean of the PGV of the two horizontal components of the ith ground motion in the set. Table 3 also shows the normalized factors for the ground motions set.

4.4 Define intensity measure

The intensity measure (IM) is intended to characterize the strength of the ground motion record. In the literature, the IM used are peak ground acceleration (PGA), peak ground velocity (PGV), and first-mode period damped spectral acceleration $Sa(T_1)$. However, the 5% damped first mode spectral acceleration $Sa(T_1,5\%)$ is more often adapted, because it minimizes the scatter in the results and provides a complete characterization of the response without the need for magnitude or source to site distance information [30]. Hence, $Sa(T_1)$ is considered as the intensity measure in this study.

4.5 Scaling record set

For collapse evaluation, ground motions are scaled to increasing earthquake intensities for $Sa(T_1)$ ranging between 0.1g and 2g. The scaling factor is defined by the following equation:

$$SF_{ij} = \frac{Sa_j}{CF_i*Sa_j(T)},$$ (4)

where $SF_{i,j}$ is the scaling factor for the ith ground motion at the jth step in the dynamic analysis; Sa_j is the mean $Sa(T)$ of the records in the set at the jth step of the analysis and CF_i is the normalization factor of the ith record in the set. Each frame will be subjected to a set of gradually scaled earthquake records, then a nonlinear analysis will be performed to obtain the seismic response. The parameters recorded for each analysis are story drift, base reactions, confined and unconfined stresses/strains at several section in the columns and beams, and the tensile stresses in the reinforcing bars.

4.6 Developing fragility curves

Evaluating eqn (2) for a given structure requires estimating θ and β from the dynamic analysis. The lognormal distribution parameters can be determined using "Method A" approach by Porter et al. [31]

$$\theta = \frac{1}{n}\sum_{i=1}^{n} lnSa_i,$$ (5)

$$\beta = \sqrt{\frac{1}{n-1}\sum_{i=1}^{n}(lnSa_i - \theta)^2}.$$ (6)

where n is the number of ground motions considered, and Sa_i is the Sa value associated with onset of collapse for the ith ground motion.

5 RESULTS AND DISCUSSION

The fragility curves for the performance limits stated earlier are plotted in Fig. 8 with respect to $Sa(T_1)$. For each frame, three curves corresponding to each damage state are illustrated to better interpret the degree of effectiveness. All three frames have similar performance when it comes to immediate occupancy level. By observing the failure mechanism of each element in the frames throughout the dynamic analysis, the critical elements at this stage are the beams. The unconfined concrete is the first to spall in the beams of all frames; this explains the similar behavior of all frames since their horizontal elements are identical. For a life safety level, Frames 2 and 3 show a modest improvement over Frame 1 between a spectral acceleration of 0.5 g and 1.5 g. This performance level is mainly governed by confined concrete crushing in beams and reinforcement yielding in columns. The maximum drift plays a considerable role in case of SSMA–RC frames, because of the large deformation presented by SSMA bars. The outstanding performance of SSMA frames is displayed amid the collapse prevention level over a range of spectral accelerations from 0 to 3.6 g. Frame 3 surpassed Frame 2 by 5% over a small interval because of the larger drift SSMA reinforced columns present. SSMA frames are largely controlled by the maximum displacement that this smart material, by its definition, manifests.

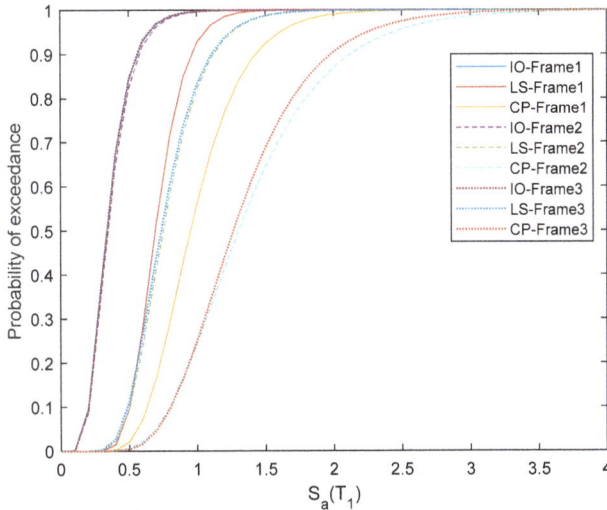

Figure 8: Fragility curves – including drifts.

Although the large deformation that an SSMA frame can undergo, as in Frame 2 for example, the vertical elements that constitute the essential components of failure are considered safe. This interpretation is better illustrated in Fig. 9, where we assumed that the maximum drift limit is not a factor that contributes to any performance level failure.

As explained earlier, the immediate occupancy is controlled by the beam behavior, so no change is expected at this level. For the second performance level, the maximum drift is one of the factors that affected the analysis in the case of SSMA–RC frames. Frames 2 and 3 surpassed Frame 1 by 40% between 0.6 g and 2 g spectral acceleration. Up to this level, replacing steel rebars by SSMA over the columns of the first floor or the entire frame is found to be invariant. This may be explained by the fact that most plastic hinges are forming at the base level of Frame 1. By substituting the reinforcement of the first floor by SSMA rebars, the frame will need a larger force to reach its unrecoverable strain limit, hence the concrete will be subjected to minimal load until the ultimate strain is reached in SSMA.

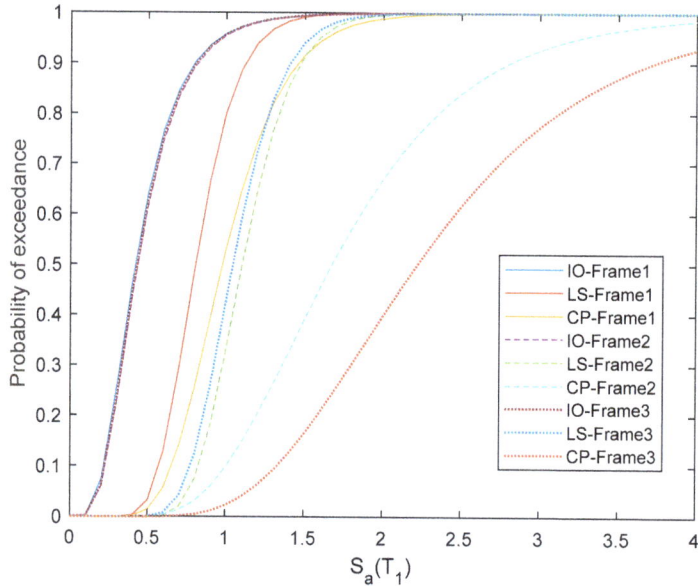

Figure 9: Fragility curves – excluding drifts.

As explained earlier, the immediate occupancy is controlled by the beam behavior, so no change is expected at this level. For the second performance level, the maximum drift is one of the factors that affected the analysis in the case of SSMA–RC frames. Frames 2 and 3 surpassed Frame 1 by 40% between 0.6 g and 2 g spectral acceleration. Up to this level, replacing steel rebars by SSMA over the columns of the first floor or the entire frame is found to be invariant. This may be explained by the fact that most plastic hinges are forming at the base level of Frame 1. By substituting the reinforcement of the first floor by SSMA rebars, the frame will need a larger force to reach its unrecoverable strain limit, hence the concrete will be subjected to minimal load until the ultimate strain is reached in SSMA.

Accordingly, confined concrete stress will be reached in Frame 1 before Frames 2 and 3. It is important to mention that the principal cause for Frame 2 to have a slightly improved performance over Frame 3 is its ability to recover plastic strains. That is to say that having a larger number of SSMA will lead to high plastic deformations, hence higher residuals. By omitting the effect of maximum displacement, it is shown that Frame 3 has the lower failure level as compared to Frames 1 and 2. Having SSMA as reinforcing bars in all the columns (Frame 3), lead to a reduction in the number plastic hinges and hence increased the overall performance. It may not be needed to have SSMA in all columns since the accumulation of plastic hinges will be essentially in the bottom floors. On that account, further SSMA configurations need to be considered to achieve the best performance level with the minimum material cost.

6 CONCLUSIONS

The aim of this study is to demonstrate the advantages of using SSMA in RC frames for different performance levels, as compared to regular steel RC frame. Since the base columns are more susceptible to damage during earthquake, the first SSMA–RC frame was reinforced by SSMA bars at the base columns. To account for additional plastic hinge formation during severe earthquakes, another RC frame reinforced with SSMA along all its columns, is used. The seismic performance of the three frames was compared using the fragility analysis. The frames were subjected to 21 incrementally scaled ground motions, where the stresses/strains in columns and beams were recorded, in addition to the drift and residual displacement. The results of the analysis are as follow:

- SSMA requires larger tensile forces to reach plastic strains due to the lower stiffness of this material; hence higher ground motion intensities are needed to reach failure.
- For an immediate occupancy performance level, steel frame showed similar behavior to both SSMA frames since the beams were the key elements at this stage. An addition frame having its beams reinforced with SSMA bars could be useful for further investigations.
- The number of SSMA bars is crucial, as the larger the steel to SSMA ratio is, the higher the deformations are. Thereby, the recovery capacity will be reduced, and the repair cost will increase.
- Since SSMA is the slowest to reach unrecoverable strain limit, it will attract both tensile and compressive stresses generated by the ground excitation, therefore the concrete part will be subjected to negligeable stresses, and the concrete failure will be delayed.

The maximum displacement of an RC frame can be dismissed from the failure criteria of a specific performance level since it may be a misleading indicator of failure; specially in case of SSMA–RC frames. The fragility analysis highlighted the efficiency of using SSMA in RC frames for both life safety and collapse prevention performance levels. The number of SSMA bars to be used and their locations are concerns to be addressed in further studies, for economical and structural purposes.

REFERENCES

[1] Takagi, T., A concept of intelligent materials. *Journal of Intelligent Material Systems and Structures,* **1**(2), pp. 149–156, 1990. DOI: 10.1177/1045389X9000100201.

[2] Reece, P., *Smart Materials and Structures: New Research*, Nova Science Publishers: New York, 2007.

[3] Mehrpouya, M. & Bidsorkhi, H., MEMS applications of NiTi based shape memory alloys: A review. *Micro and Nanosystems,* **8**(2), pp. 79–91, 2017. DOI: 10.2174/1876402908666161102151453.

[4] Mir, B.A., Smart materials and their applications in civil engineering: An overview. *International Journal of Civil Engineering and Construction Science*, pp. 11–20, 2017.

[5] Murty, C.V.R., https://www.iitk.ac.in/nicee/EQTips/EQTip17.pdf.

[6] Molod, M.A., Spyridia, P. & Barthold, F-J., Applications of shape memory alloys in structural engineering with a focus on concrete construction: A comprehensive review. *Construction and Building Materials*, **337**, 127565, 2022. DOI: 10.1016/j.conbuildmat.2022.127565.

[7] Hong, C., Qian, H. & Song, G., Uniaxial compressive behavior of concrete columns confined with superelastic shape memory alloy wires. *Materials*, **13**(5), p. 1227, 2020. DOI: 10.3390/ma13051227.

[8]   Hamid, N.A., Ibrahim, A. & Adnan, A., Behaviour of smart reinforced concrete beam with super elastic shape memory alloy subjected to monotonic loading. *AIP Conference Proceedings*, **1958**, 020034, 2018. DOI: 10.1063/1.5034565.
[9]   Khaloo, A.R., Eshghi, I. & Aghl, P.P., Study of behavior of reinforced concrete beams with smart rebars using finite element modeling. *International Journal of Civil Engineering*, **8**(3), 2010.
[10]   Abdulridha, A., Palermo, D., Foo, S. & Vecchio, F.J., Behavior and modeling of superelastic shape memory alloy reinforced concrete beams. *Engineering Structures,* pp. 893–904, 2013. DOI: 10.1016/j.engstruct.2012.12.041.
[11]   Saiidi, M.S. & Wang, H., Exploratory study of seismic response of concrete columns with shape memory alloys reinforcement. *ACI Structural Journal*, **103**(3), pp. 436–443, 2006. DOI: 10.14359/15322.
[12]   Alam, S., Youssef, A. & Nehdi, M., Seismic behaviour of concrete beam-column joints reinforced with superelastic shape memory alloys. *9th Canadian Conference on Earthquake Engineering,* Canada, 2007. DOI: 10.13140/2.1.4516.0966.
[13]   Nahar, M., Muntasir Billah, A.H.M., Kamal, H.R. & Islam, K., Numerical seismic performance evaluation of concrete beam-column joint reinforced with different super elastic shape memory alloy rebars. *Engineering Structures*, **194**, pp. 161–172, 2019. DOI: 10.1016/j.engstruct.2019.05.054.
[14]   Alam, S., Nedi, M. & Youssef, M., Seismic performance of concrete frame structures reinforced with superelastic shape memory alloys. *Smart Structures and Systems,* pp. 565–585, 2009. DOI: 10.12989/sss.2009.5.5.565.
[15]   McKenna, F., Fenves L. & Scott H., Open System for Earthquake Engineering Simulation (OpenSees). *Pacific Earthquake Engineering Research Center*, 2000.
[16]   ACI 318-19, Building code requirements for structural concrete. American Concrete Institute: Farmington Hills, MI, USA, 2020.
[17]   ASCE/SEI 7-16, Minimum design loads and associated criteria for buildings and other structures, American Society of Civil Engineers, 2017.
[18]   IBC, International building code, Virginia, 2020.
[19]   Aarango, M. & Lubkowski Z., Seismic hazard assessment and design requirements for Beirut, Lebanon. *15th World Conference in Earthquake Engineering,* Lisbon, 2012.
[20]   SAP2000, Integrated structural analysis and design software. *Computer and Structures*, 1997.
[21]   Menegotto, M. & Pinto, P.E., Method of analysis of cyclically RC plane frames including changes in geometry and non-elastic behavior of elements under normal force and bending. *Materials Science*, pp. 15–22, 1973. DOI: 10.5169/seals-13741.
[22]   Anggraini, R., Tavio, I., Raka, G.P. & Agustiar, Stress-strain relationship of high-strength steel (HSS) reinforcing bars. *AIP Conference Proceedings*, **1964**, 020025, 2018. DOI: 10.1063/1.5034565.
[23]   Popovics, S., A numerical approach to the complete stress strain curve for concrete. *Cement and Concrete Research,* pp. 583–599, 1973. DOI: 10.1016/0008-8846(73)90096-3.
[24]   Karsan, D. & Jirsa O., Behavior of concrete under compressive loading. *Journal of Structural Division ASCE*, 1969. DOI: 10.1061/JSDEAG.0002424.
[25]   Scott, M.H., A tale of two element formulations. Portwood Digital, 2020. https://portwooddigital.com/2020/02/23/a-tale-of-two-element-formulations/.
[26]   FEMA-356, *Handbook for the Seismic Evaluation of Buildings: A Prestandard and Commentary for Seismic Rehabilitation of Buildings*, The American Society of Civil Engineers for the Federal Emergency Management Agency: Washington, DC, 2000.

[27] Padgett, J. & Desroches, R., Sensitivity of seismic response and fragility to parameter uncertainty. *Journal of Structural Engineering*, **133**(12), 2007. DOI: 10.1061/(asce)0733-9445(2007)133:12(1710).

[28] Asgarian, B., Salehi, E. & Shokrgozar, H., Probabilistic seismic evaluation of buckling restrained braced frames using DCFD and PSDA methods. *Earthquakes and Structures*, **10**(1), pp. 1–19, 2016. DOI: 10.12989/eas.2016.10.1.105.

[29] PEER Ground Motion Database, University of California, Berkeley, 2014.

[30] Carballo, J.E. & Cornell, C.A., Probabilistic seismic demand analysis: Spectrum matching and design. Reliability of marine structures program, Report No. RMS-41, Stanford University, USA, 2000.

[31] Porter, K., Kennedy, R. & Bachman R., Creating fragility functions for performance-based earthquake engineering. *Earhquake Spectra*, pp. 471–489, 2007. DOI: 10.1193/1.2720892.

Author index

WITPRESS ...for scientists by scientists

Structural Studies, Repairs and Maintenance of Heritage Architecture XVII & Earthquake Resistant Engineering Structures XIII

Edited by: **S. HERNÁNDEZ,** *University of A Coruña, Spain and* **G. MARSEGLIA,** *Link Campus University, Italy*

Structural Studies, Repairs and Maintenance of Heritage Architecture XVII

The importance of retaining the built cultural heritage cannot be overstated. Rapid development and inappropriate conservation techniques are threatening many heritage unique sites in different parts of the world.

Selected papers presented at the 17th International Conference on Studies, Repairs and Maintenance of Heritage Architecture are included in this volume. They address a series of topics related to the historical aspects and the reuse of heritage buildings, as well as technical issues on the structural integrity of different types of buildings, such as those constructed with materials as varied as iron and steel, concrete, masonry, wood or earth.

Contributions originate from scientists, architects, engineers and restoration experts from all over the world and deal with different aspects of heritage buildings, including how to formulate regulatory policies, to ensure effective ways of preserving the architectural heritage.

Earthquake Resistant Engineering Structures XIII

Papers presented at the 13th International Conference on Earthquake Resistant Engineering Structures form this volume and cover basic and applied research in the various fields of earthquake engineering relevant to the design of structures.

Major earthquakes and associated effects such as tsunamis continue to stress the need to carry out more research on those topics. The problems will intensify as population pressure results in buildings in regions of high seismic vulnerability. A better understanding of these phenomena is required to design earthquake resistant structures and to carry out risk assessments and vulnerability studies.

The problem of protecting the built environment in earthquake-prone regions involves not only the optimal design and construction of new facilities but also the upgrading and rehabilitation of existing structures including heritage buildings. The type of highly specialized retrofitting employed to protect the built heritage is an important area of research.

The included papers cover such topics as Seismic hazard and tsunamis; Building performance during earthquakes; Structural vulnerability; Seismic isolation and energy dissipation; Passive earthquake protection systems.

ISBN: 978-1-78466-429-9   **eISBN: 978-1-78466-430-5**
Published 2021 / 444pp

www.ingramcontent.com/pod-product-compliance
Lightning Source LLC
Chambersburg PA
CBHW062002190326
41458CB00009B/2942